About the NCCN Guidelines for Patients®

National Comprehensive Cancer Network®

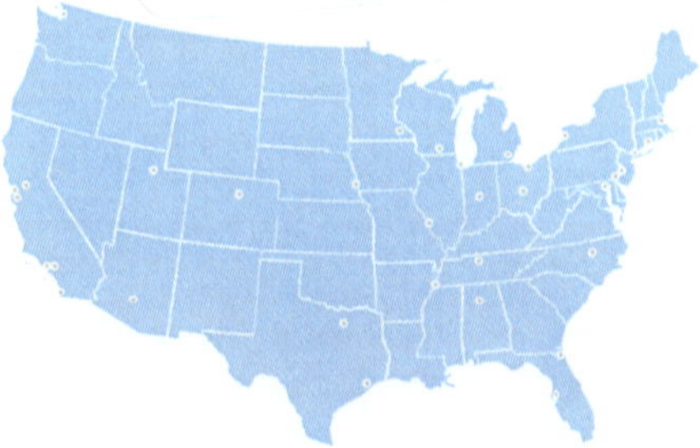

Did you know that top cancer centers across the United States work together to improve cancer care? This alliance of leading cancer centers is called the National Comprehensive Cancer Network® (NCCN®).

Cancer care is always changing. NCCN develops evidence-based cancer care recommendations used by health care providers worldwide. These frequently updated recommendations are the NCCN Clinical Practice Guidelines in Oncology (NCCN Guidelines®). The NCCN Guidelines for Patients plainly explain these expert recommendations for people with cancer and caregivers.

These NCCN Guidelines for Patients are based on the NCCN Clinical Practice Guidelines in Oncology (NCCN Guidelines®) for Genetic/Familial High-Risk Assessment: Breast, Ovarian, Pancreatic, and Prostate, Version 2.2025 – November 7, 2024

View the NCCN Guidelines for Patients free online
NCCN.org/patientguidelines

Find an NCCN Cancer Center near you
NCCN.org/cancercenters

Connect with us

Supporters

NATIONAL COMPREHENSIVE CANCER NETWORK®
NCCN FOUNDATION
Guiding Treatment. Changing Lives.

NCCN Guidelines for Patients are supported by funding from the NCCN Foundation®

NCCN Foundation gratefully acknowledges the following corporate supporters for helping to make available these NCCN Guidelines for Patients: AstraZeneca; GSK; and Ipsen Biopharmaceuticals, Inc.

NCCN independently adapts, updates, and hosts the NCCN Guidelines for Patients. Our corporate supporters do not participate in the development of the NCCN Guidelines for Patients and are not responsible for the content and recommendations contained therein.

To make a gift or learn more, visit online or email

NCCNFoundation.org/donate

PatientGuidelines@NCCN.org

Contents

NCCN Foundation seeks to support the millions of patients and their families affected by a cancer diagnosis by funding and distributing NCCN Guidelines for Patients. NCCN Foundation is also committed to advancing cancer treatment by funding the nation's promising doctors at the center of innovation in cancer research. For more details and the full library of patient and caregiver resources, visit NCCN.org/patients.

National Comprehensive Cancer Network (NCCN) and NCCN Foundation
3025 Chemical Road, Suite 100, Plymouth Meeting, PA 19462 USA

1
About genetic testing

If you've been diagnosed with cancer, genetic testing can help your oncologist make the most informed recommendations for your care. If you don't have cancer but it runs in your family, genetic testing can help you understand your chance of developing it and take steps to lower your risk.

Which hereditary cancers are discussed in this guide?

This resource addresses testing for inherited gene variants (differences) associated with the following cancers and cancer syndromes:

- Breast cancer
- Ovarian cancer
- Pancreatic cancer
- Prostate cancer
- Li-Fraumeni syndrome (LFS)
- Cowden syndrome/*PTEN* hamartoma tumor syndrome

For information on genetic testing for other hereditary cancers and syndromes, health care providers are encouraged to refer to the *NCCN Clinical Practice Guidelines in Oncology for Genetic/Familial High-Risk Assessment: Colorectal, Endometrial, and Gastric*.

Hereditary cancers aren't very different from non-hereditary cancers. It's more the pattern of how they occur in your family that makes them different. Hereditary cancers often develop earlier and affect multiple relatives on the same side of the family.

What causes hereditary cancers?

Hereditary cancers are caused by inherited variants (differences) in specific genes. Together these genes are called cancer predisposition (or cancer susceptibility) genes.

Gene variants you are born with are called germline. They are part of the DNA of every cell in your body. In addition to your DNA, your risk for cancer is based on your personal and family health history.

What is penetrance?

In someone with a gene variant that increases the risk of a health problem, penetrance describes how likely they are to develop symptoms or features of the problem. For example, 100% penetrance means that everyone with that variant will develop the problem if they live long enough. This is very rare.

Cancer predisposition genes are described as low-, moderate-, or high-penetrance. Cancer-causing variants in high-penetrance genes are generally more likely to result in a cancer diagnosis than variants in low- or moderate-penetrance genes. But other factors also affect how likely you are to develop cancer, including your age, family history, environment, and lifestyle.

Why you should read this book

Making decisions about cancer care can be stressful. You may need to make tough decisions under pressure about complex choices.

NCCN Guidelines for Patients are trusted by patients and providers. They clearly explain current care recommendations made by respected experts in the field. Recommendations are based on the latest research and practices at leading cancer centers.

Cancer care is not the same for everyone. By following expert recommendations for your situation, you are more likely to improve your care and have better outcomes as a result. Use this book as your guide to find the information you need to make important decisions.

2
The genetic testing process

This chapter describes what's involved in determining if you meet the criteria for genetic testing, and what to expect after a positive result.

Determining your risk for hereditary cancer and learning about genetic testing involves 3 steps:

- Counseling before any testing is ordered
- Choosing the best test(s) for you
- Counseling when your results are ready

Ideally, a health care professional trained in cancer genetics will be involved at each stage. This might be a genetic counselor, clinical geneticist, oncologist, oncology nurse, or other health care provider.

Pre-test counseling

Before any testing is ordered, it's important to learn what the results could mean for you and your family. A genetics professional or other provider will explain the benefits, risks, and limitations of genetic testing for cancer risk.

Your health history

Your past and current health is a key factor in determining your risk of hereditary cancer. If you've been diagnosed with any cancers, your genetics provider will ask about:

- the type(s) of cancer
- your age at diagnosis
- what cancer treatments you had (some can cause late side effects)

Whether you've had any surgeries to lower your risk of developing cancer is also important to share. Examples include risk-

The genetic testing process

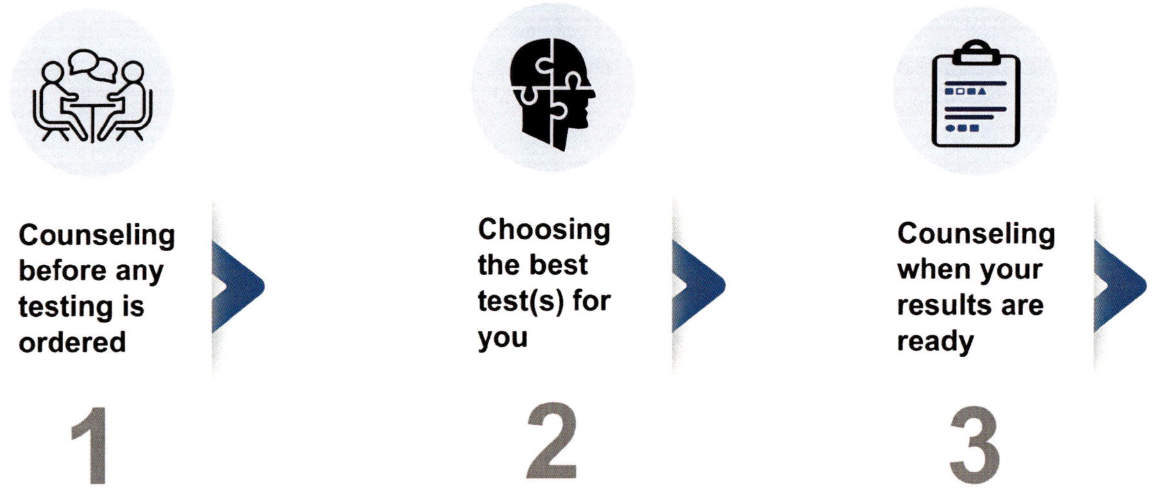

reducing salpingo-oophorectomy (RRSO) and risk-reducing mastectomy (RRM). RRSO removes both ovaries and fallopian tubes. This surgery is also called a bilateral salpingo-oophorectomy (BSO). RRM removes both breasts.

Your provider is also likely to ask about hormonal factors that may affect your risk, such as:

- your pregnancy history
- whether you've used birth control pills or other hormonal birth control
- whether you've received menopausal hormone therapy

Your family health history

Your testing provider will collect detailed information about the health of your family members.

Whether you have any close relatives with cancer is especially important. Close blood relatives include first-, second-, and third-degree relatives on each side of your family. See the chart on the next page for details.

The type(s) of cancer your relatives were diagnosed with, their age at diagnosis, and the cancer laterality is also important. In paired organs, like the breasts and ovaries, laterality describes which of the paired organ(s) have cancer.

They will also ask about your ethnicity. Founder variants are cancer-causing variants found in a specific group of people that can be traced back to common ancestors. People of Ashkenazi Jewish ancestry and other groups are at increased risk of founder variants, including *BRCA1/2* variants.

Prior hematopoietic cell transplant

If you've had an allogeneic hematopoietic cell transplant (HCT) to treat blood cancer, your genetic testing results could reflect the DNA of the stem cell *donor* when using blood, saliva, or a cheek swab. To make sure your DNA (and not your donor's) is extracted, your provider may remove a small piece of your skin or other tissue for testing.

How much does testing cost?

Ask your provider about the cost of genetic testing. Many insurance companies will cover the cost if a health care provider confirms it's needed for your health care. In the United States, most people can get genetic testing for $250 or less out-of-pocket, depending on the number of genes tested and other factors.

Other things to discuss

Other points you and your testing provider will discuss include:

- Informed consent
- The privacy of your genetic information
- Current laws on genetic discrimination

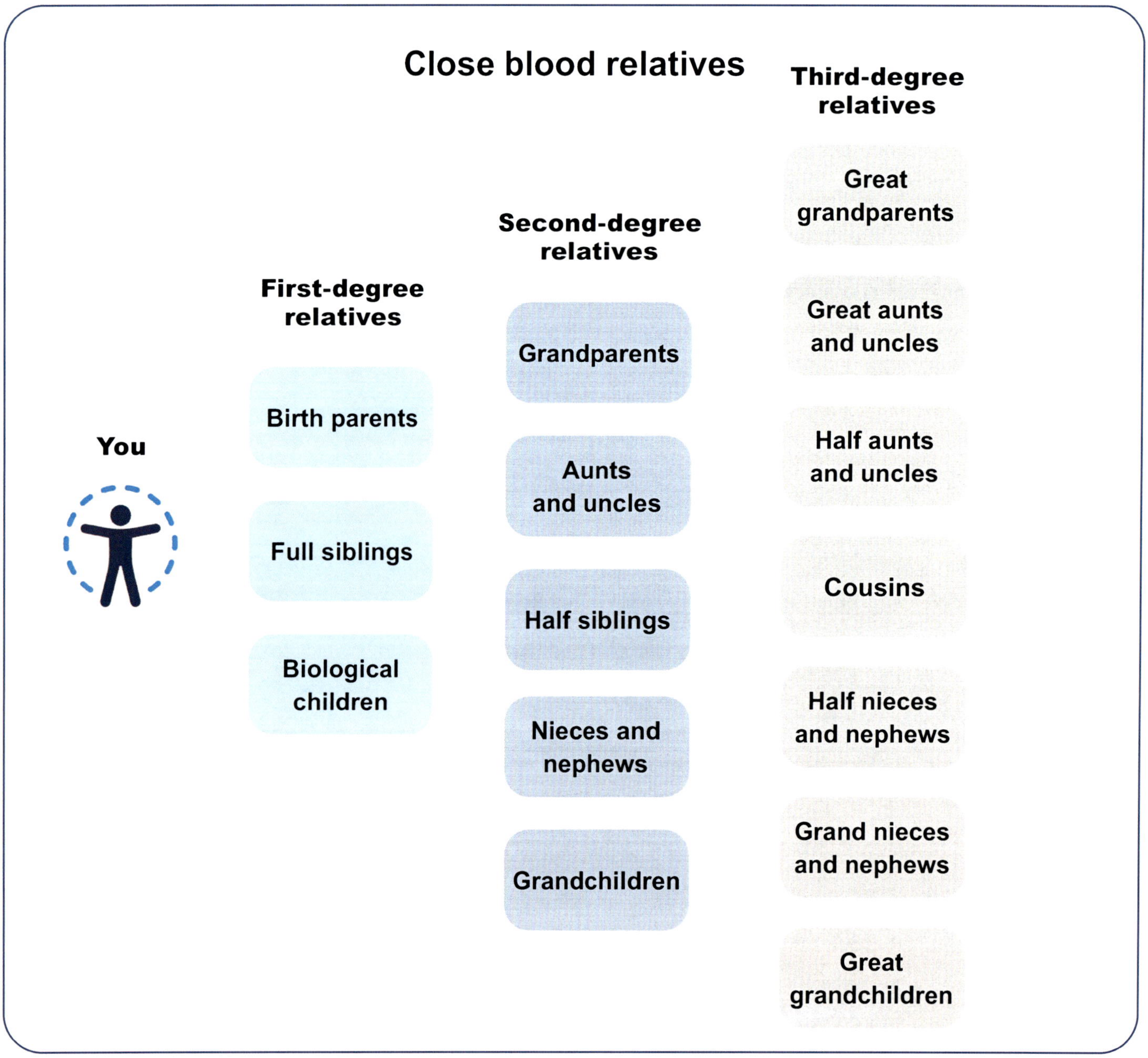

Choosing the best test(s) for you

Genetic testing is often performed using next-generation sequencing (NGS). Also called a multi-gene panel, NGS analyzes entire sets of cancer predisposition genes at the same time. Multi-gene panels may include both moderate- and high-penetrance genes.

To be tested, you will typically provide a sample of blood or saliva (spit). The sample is sent to a lab to check for gene variants that cause hereditary cancers and syndromes.

Other types of samples may be used to obtain your DNA if needed. Testing can be performed on a sample of your cheek cells (buccal smear), skin, hair, or nails.

At-home genetic tests

Genetic testing offered by companies that provide ancestry and health information, like 23andMe, is called recreational testing. Some downsides of these tests include:

- They only check for variants linked with a few cancers
- They don't consider your family history
- Results can be inaccurate

If you had a positive result from a recreational test, discuss the result with a health care provider. You may need to do another genetic test to confirm the results before using them for your health care.

More comprehensive at-home tests are available through companies like Invitae and Color Genomics. While these typically include the services of a genetic counselor, they often provide only limited opportunity to discuss and understand the significance of a gene test result.

If results are positive

After testing, your genetics provider will explain your results and what they mean for your health and cancer risk.

A positive result means that a pathogenic (cancer-causing) or a likely pathogenic variant was found in 1 or more of the genes included in the testing panel. Seeing a genetics provider is recommended for positive results.

For those who test positive and don't have a cancer diagnosis, the next step is discussing options for managing your cancer risk with your provider. This could involve more frequent screening, risk-reducing medicines, or risk-reducing surgery.

Some medical centers include specialized care and screening services for people with a positive genetic testing result. If available, your provider may refer you to these services. Ask your provider about cancer prevention clinics, hereditary cancer support or advocacy groups, and any available research studies.

If you are being treated for cancer, a positive result may affect your cancer care. It may also affect the clinical trials you are eligible for.

Sharing your results

To the extent possible, do your best to share your results with your biological family members. They may be at increased risk as well and can benefit from risk assessment and possibly testing.

It's helpful if you provide them with a copy of your results that they can share with their doctors and genetics professional. Some providers offer easy-to-understand materials and resources you can use to notify your family about your results and their risk.

Referral to a genetics provider

If your testing location doesn't have the resources for pre-test counseling or post-test care, ask to be referred to a genetics provider.

Seeing a genetics provider is also recommended for a positive result, and for a complex result that needs expert interpretation and management.

Re-consultation about your risk

Over time, it may be helpful to have a follow-up visit with a genetics professional or other health care provider who is familiar with inherited cancer risk.

This visit allows you to check in with your provider about your cancer screening strategy. They may suggest changes based on the latest research, or if any new cancers have been diagnosed in your family. You can also discuss any newer options for genetic testing.

And if your life circumstances have changed, this is an opportunity to revisit earlier decisions about risk-reducing surgery, and other screening and prevention choices.

How often you should have follow-up depends on your age, reproductive planning, other health problems, risk-reducing surgeries, and other risk factors.

Family planning

If you are of reproductive age, your provider will explain what a positive genetic testing result could mean for starting or growing your family.

Some couples who have or carry genetic disorders choose to conceive naturally. Between the 10th and 14th weeks of pregnancy, the baby can be tested for cancer-causing variants using chorionic villus sampling (CVS). This is called prenatal diagnosis. Depending on the results, the parents may make difficult decisions regarding pregnancy management.

Another option for couples who have or carry genetic disorders is in vitro fertilization (IVF). In IVF, eggs are fertilized with sperm in a lab to create embryos. The embryos are implanted into the uterus or frozen for future use. Before being transferred to the uterus, the embryos can be tested for hereditary cancer-causing variants. This is called pre-implantation genetic testing (PGT). If desired, only embryos not carrying the variant(s) can be transferred to the uterus.

Couples can also choose to use eggs, sperm, or embryos from a donor for IVF.

Pre-implantation genetic testing (PGT)

Embryos created through IVF can be tested for a hereditary cancer risk variant known to be carried by the egg or sperm donor. If desired, only embryos not carrying the variant can be transferred to the uterus.

Other possible results

Other possible results of genetic testing are described next.

A negative result means that you don't carry a pathogenic variant in the cancer-predisposing genes included in the testing panel. A negative result is either a true negative or an uninformative negative.

True negative results

If a cancer-causing variant runs in your family and you don't carry it, it's called a true negative. This is a reassuring result.

Keep in mind that you may still be at higher risk than average based on your personal and family history. You may still be at risk for any of the following reasons:

- You carry a variant in a tested gene that can't be detected using current technology
- You carry a variant in a gene that wasn't included in the testing panel
- You have several family members diagnosed with cancer, but none is linked to a known hereditary cancer gene

Uninformative negative results

If you test negative, and your relatives with early-onset or suspicious cancers haven't been tested, it's called an uninformative negative.

It's uninformative because you still don't know whether they carried a variant that you didn't inherit, or whether their cancers can't be explained by known hereditary cancer genes.

Even though you tested negative, you may benefit from increased cancer screening and risk-reducing measures based on your family history. Some medical centers include specialized high-risk clinics that offer this type of family history-based screening.

Your family members may benefit from genetic testing, both to learn their own cancer risk and to contribute data about your family's overall cancer risk.

Over time, you may be a candidate for additional genetic testing due to new family history data and advances in testing.

Variant of uncertain significance

A variant of uncertain significance (VUS) is a genetic variant for which little information is available. This is a common result, especially when a large number of genes are tested at the same time.

Over time, more information may become available and your health care provider may provide you with an update on the meaning of these variants. Until then, these aren't used to make care decisions.

Mosaic results

When someone has two or more sets of cells that differ genetically from one another, the results are called mosaic. This is thought to be caused by an error in mitosis (cell division) in a growing baby during pregnancy.

Mosaicism can lead to many kinds of disorders. Referral to a genetics provider is recommended for mosaic results.

Key points

- Determining your risk for hereditary cancer involves counseling before testing, choosing the best gene(s) to test, and counseling when your results are ready.
- Your health and cancer history, and that of your biological family members, are key factors in determining your risk of hereditary cancer.
- Genetic testing is often performed using next-generation sequencing (NGS). To be tested, you will typically provide a sample of blood or saliva.
- If a cancer-causing variant is found, share your results with your biological family members. They may also be at increased risk and can benefit from risk assessment and possibly testing.
- Family planning options for couples who have or carry genetic disorders include prenatal diagnosis after natural conception, and in vitro fertilization (IVF) with pre-implantation genetic testing (PGT).
- Other possible results include true negative, uninformative negative, variant of uncertain significance, and mosaic results. If results are negative, you may still be at higher risk than average based on your family history.
- Over time, check in with your genetics provider about your cancer screening and prevention strategy. Tell them about any new cancers diagnosed in your family and learn about any newer options for testing.

Questions to ask

- What are the advantages and disadvantages to having genetic testing?
- How long does it take to get the results?
- Can my health or life insurance company increase my rates if I test positive?
- I don't think I can handle the emotional impact of a positive result. What do I do?
- What can I do to lower my risk if I test positive?

3

Testing criteria and results-based care

This chapter describes the criteria used to determine whether genetic testing would be helpful for you, as well as recommended care for a positive result.

General recommendations

Testing is recommended in the situations described on this page. If you have a personal or family history of breast, ovarian, pancreatic, or prostate cancer, more specific guidance is provided later in this chapter.

Cancer-causing variants in relatives

If you have any blood relatives with a confirmed cancer-causing gene variant, genetic testing is recommended.

To help make treatment decisions

If you have cancer, genetic testing can help your provider make decisions about chemotherapy and other forms of systemic therapy. Specifically, whether platinum chemotherapy and PARP inhibitors would be helpful for certain cancers.

The results can also guide decisions about whether to have risk-reducing surgery in order to prevent breast or ovarian cancer.

The cancer has a mutation

If you have cancer, your tumor tissue may have been tested for features, called biomarkers, that can help guide your treatment. Biomarker testing is different than genetic testing for germline (inherited) mutations. Biomarkers are often mutations (changes) in particular genes. Mutations in the tumor or cancer itself are called somatic, acquired, or simply tumor mutations.

For some hereditary cancers, planning the best treatment depends on whether you were born with the same difference in DNA that was found in your tumor. If the cancer itself has a pathogenic or likely pathogenic (P/LP) mutation and it would impact your care if the variant is also germline (inherited), genetic testing is recommended.

Who else might benefit from testing?

People of **Ashkenazi Jewish ancestry** are at higher risk than average of carrying certain cancer-causing gene variants. Testing may be considered if you are of Ashkenazi Jewish ancestry, even if you don't have any other risk factors.

Genetic testing may also be considered for those diagnosed with **serous endometrial cancer**, a rare type of uterine cancer that may be linked with *BRCA* variants.

Breast cancer

Breast cancer starts in the cells of the breast. Almost all breast cancers are carcinomas. Carcinomas are cancers that start in the cells that line the inner or outer surfaces of the body. Anyone can develop breast cancer, including those assigned male at birth.

There are different types of breast carcinoma. The most common types are either ductal or lobular.

- **Ductal carcinoma** starts in the cells that line the milk ducts. Ductal carcinoma is the most common type of breast cancer.
- **Lobular carcinoma** starts in the lobules (milk glands) of the breast.

In this patient guide, the term breast cancer includes invasive ductal carcinoma (IDC) and ductal carcinoma in situ (DCIS).

A hereditary cause is identified in about 1 in 10 breast cancers. Breast cancer susceptibility genes include *BARD1, BRCA1, BRCA2, CDH1, CHEK2, NF1, PALB2, PTEN, STK11, and TP53.*

If you meet the testing criteria described next, personalized risk assessment and genetic counseling are recommended. Genetic testing is often part of the process.

Managing your risk

In people at increased risk of hereditary breast cancer, screening with mammograms, breast MRIs, or both is often recommended in order to detect cancer growth early.

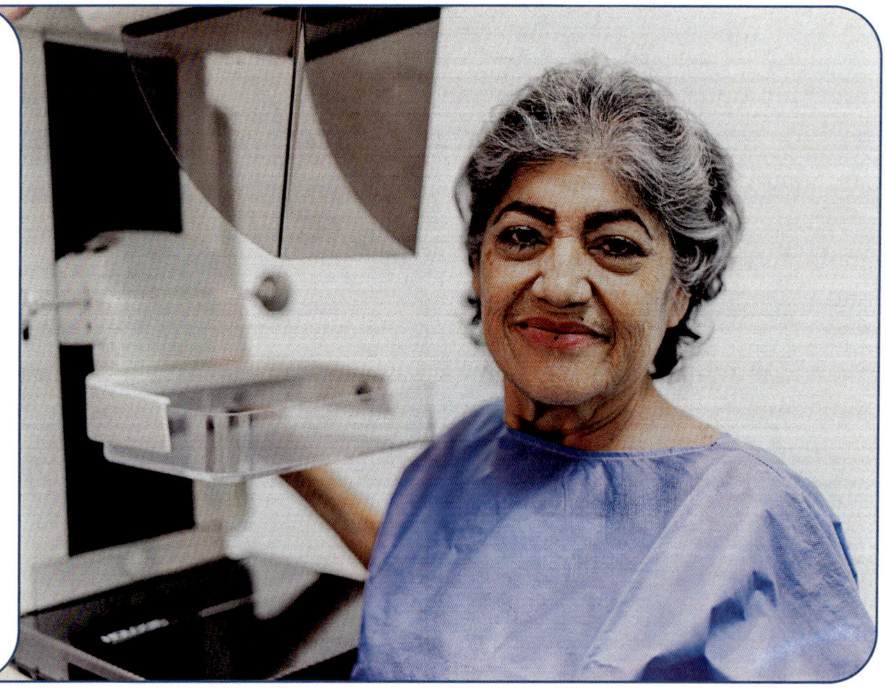

Testing in people with breast cancer

If you've been diagnosed with breast cancer and any of the following apply to you, genetic testing is recommended.

Age – You were diagnosed with breast cancer at or before age 50

Ancestry – You are of Ashkenazi Jewish ancestry

Cancer type – You've been diagnosed with any of the following:

- Male breast cancer
- Triple-negative breast cancer
- More than 1 type of breast cancer
- Lobular breast cancer and you have a personal or family history of diffuse gastric cancer

Family history – You have 1 or more close blood relatives with any of the following:

- breast cancer diagnosed at age 50 or earlier
- male breast cancer
- ovarian cancer
- pancreatic cancer
- metastatic or high-risk prostate cancer

Testing is also recommended if you have 3 relatives (including you) on the same side of your family that have breast or prostate cancer.

To guide your care – In people with metastatic breast cancer, genetic testing is recommended to help decide whether PARP inhibitor therapy should be considered.

In those with high-risk, HER2-negative breast cancer, testing is recommended to help decide if PARP inhibitor therapy would be helpful after primary treatment. Ask your oncologist if the cancer is high risk.

If you are receiving treatment for any type of cancer, a positive result may affect the clinical trials you are eligible for.

If you don't meet the criteria just described

If you haven't been diagnosed with breast cancer, or you have but don't meet the criteria just described, testing **is recommended** in the following situations.

- You have a first- or second-degree blood relative that meets the criteria just described
- You have a 5% or higher risk of a *BRCA1/2* variant based on an earlier risk calculator, like BRCAPro and CanRisk.

Genetic testing may also be **considered** if any of the following apply to you:

- You were diagnosed with breast cancer between age 51 and age 65.
- You were diagnosed with breast cancer (at any age) and have 1 or more close blood relatives with certain types of prostate cancer.
- You have a history of malignant phyllodes tumors.

Testing may also be an option for those with a 2.5% to 5% risk of a *BRCA1/2* variant, according to an earlier risk calculator. In this case, when deciding whether to include genes other than *BRCA1/2*, your provider will consider your preferences, your insurance coverage, and the speed of results. The more genes that are tested, the more likely you are to get a variant of uncertain significance (VUS).

Who isn't likely to benefit from testing?

In people who **don't have** close relatives with breast, ovarian, pancreatic, or prostate cancer, testing is unlikely to find high-risk variants in the following groups:

- Those assigned female at birth diagnosed with breast cancer after age 65
- Men diagnosed with localized prostate cancer and a Gleason Score below 7. The Gleason score usually ranges from 6 to 10. The lower the score, the slower the cancer cells are expected to grow and spread.

Managing your breast cancer risk after a positive result

A positive result means that a germline pathogenic or likely pathogenic variant was found in the susceptibility genes included in the testing panel.

Based on your results, your provider may recommend some of the strategies described next to reduce your risk of breast cancer.

Recommendations for lowering risk are guided by the penetrance of the specific gene variant(s) identified. See **Guide 8** on page 35 for complete guidance according to gene.

Breast awareness

Breast awareness involves becoming familiar with your breasts and reporting any changes to your health care provider. Doing breast self exams on a regular basis may help you notice changes. If you are premenopausal, doing the exams when your period is almost over may be most helpful.

Screening mammogram

A screening mammogram uses low-dose x-rays to take 4 pictures of your breast (two on each side). A radiologist reviews the images and decides if you need additional imaging. If you do, you will likely have a diagnostic mammogram.

A diagnostic mammogram usually involves taking more detailed x-ray pictures of the breast from different angles to check the suspicious area more closely.

Clinical breast exam

Clinical breast exam (CBE) is a physical exam of the bare breast performed by a health care provider to check for lumps or other changes. While you are seated or lying down, your provider will palpate (feel) the entire breast, including the armpit. A nurse or assistant might also be in the room during the exam.

Breast MRI

A magnetic resonance imaging (MRI) scan uses radio waves and powerful magnets to take pictures of areas inside the body. It doesn't use radiation.

Your provider may recommend breast MRI in addition to mammography. They will consider your age, family history, breast density, and preferences to determine whether to use both methods.

For a breast MRI, a gadolinium-based contrast agent (GBCA)—a rare, heavy metal—is used to enhance the quality of the MRI. There are no harmful effects from GBCA, but it may linger in the body for months to years afterward. Talk to your provider if you have any concerns. The MRI should be done with and without contrast.

Risk-reducing mastectomy

Having surgery to remove the breasts lowers the risk of developing breast cancer. Surgery to remove both breasts for this purpose is called risk-reducing bilateral mastectomy.

Risk-reducing mastectomy is an option to consider for those with germline variants in certain breast cancer susceptibility genes, including *BRCA1/2*, *CDH1*, *PALB2*, *PTEN* (associated with Cowden syndrome), *STK11*, and *TP53* (associated with Li-Fraumeni

Breast MRI

For those at increased risk of breast cancer, breast MRI may be recommended in addition to mammography. Your provider will consider your specific gene variant(s), age, family history, breast density, and preferences to determine whether to use both methods.

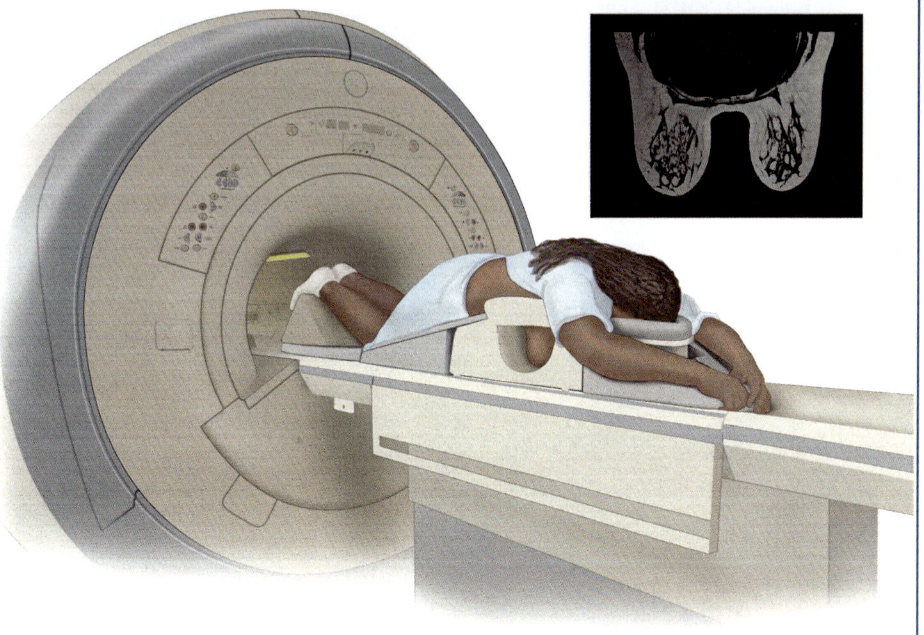

syndrome). Care for Li-Fraumeni syndrome (page 29) and Cowden syndrome (page 32) is provided later in this chapter.

Your provider will explain the risks of this surgery, the extent of cancer protection it provides, and your options for breast reconstruction. If you have a strong family history of breast cancer, take this into consideration when making decisions about risk-reducing mastectomy.

Having your breasts removed can have a major impact on your emotional well-being and quality of life. Ask your provider about resources and information available to you.

Risk-reducing drugs

Certain medicines can help lower the risk of developing breast cancer in those at high risk.

In **premenopausal** people, tamoxifen is most commonly used. This selective estrogen receptor modulator (SERM) reduces the amount of estrogen in the body.

Tamoxifen is typically prescribed as a 20 mg pill for 5 years. Sometimes it is prescribed for fewer years at a lower dose.

In **postmenopausal** people, either raloxifene (Evista) or an aromatase inhibitor is typically recommended.

Raloxifene is an estrogen blocker like tamoxifen. For breast cancer prevention, raloxifene is typically prescribed at a dose of 60 mg per day for 5 years.

Anastrozole and exemestane are aromatase inhibitors. These lower estrogen levels by stopping aromatase (an enzyme in fat tissue) from converting other hormones in the body into estrogen.

Let us know what you think!

Please take a moment to complete an online survey about the NCCN Guidelines for Patients.
NCCN.org/patients/response

Ovarian cancer

Most ovarian cancers actually start in the fallopian tubes. The cancer forms in the surface layer of tissue, called the epithelium. There are more than 5 types of epithelial ovarian cancer. The most common forms are:

- High-grade serous carcinoma (HGSC)
- High-grade endometrioid carcinoma

Ovarian cancer susceptibility genes include *ATM*, *BRCA1*, *BRCA2*, *BRIP1*, *PALB2*, *RAD51C*, *RAD51D*, and the Lynch syndrome genes (*MLH1*, *MSH2*, *MSH6*, *EPCAM*).

Testing criteria

Testing is recommended for anyone who has been diagnosed with epithelial ovarian cancer (including fallopian tube cancer and peritoneal cancer) at any age.

If you've never been diagnosed with ovarian cancer, testing is recommended if:

- You have a first- or second-degree blood relative diagnosed with epithelial ovarian cancer (including fallopian tube cancer or peritoneal cancer) at any age.
- You don't meet the criteria above but have a probability above 5% of a *BRCA1/2* variant based on an earlier risk calculator, like BRCAPro or CanRisk.

Based on your specific results, your provider may recommend taking steps to reduce your risk of ovarian cancer, such as risk-reducing surgery.

Risk-reducing salpingo-oophorectomy

Surgery to remove the ovaries and fallopian tubes in order to lower the risk of cancer is called risk-reducing salpingo-oophorectomy (RRSO). RRSO is recommended for those with germline cancer-causing variants in the *BRCA1/2*, *BRIP1*, *RAD51C*, and *RAD51D* genes. And it's an option for those with variants in several other genes.

RRSO can impact your health, emotional well-being, and quality of life. If you will be premenopausal when RRSO is performed, arrange to consult with a menopause specialist beforehand. More information on surgical menopause is provided on the next page.

Option of hysterectomy at time of RRSO
People with germline *BRCA* variants may be at slightly higher risk of serous uterine cancer than the average person. Ask your provider about the risks and benefits of removing your uterus at the same time as RRSO.

Surgical menopause

If you are premenopausal, surgically removing the ovaries results in a sudden drop in estrogen in the body. This drop can cause symptoms of menopause, including:

- Hot flashes
- Sleeping problems
- Night sweats
- Changes in mood
- Vaginal lining can become thin, dry, and irritated (vaginal atrophy)
- Cognitive changes
- Weight gain

There are also long-term risks of having a low estrogen level, including heart disease and weak bones (osteoporosis).

If you have symptoms, your provider may suggest menopausal hormone therapy (MHT). This approach used to be called hormone replacement therapy (HRT). It can help lessen side effects and is something to consider if you don't have breast cancer.

If your **uterus is left in place** at the time of RRSO, MHT with estrogen alone may increase your risk of endometrial cancer. For this reason, options typically include:

- Levonorgestrel intrauterine device (IUD) and estrogen given as a pill or a patch placed on the skin
- Combination estrogen with a selective estrogen receptor modulator (SERM, such as bazedoxifene)
- Combination oral contraceptive pills
- Other combinations of estrogen and progestin

Surgical menopause

In premenopausal people, surgically removing the ovaries can cause symptoms of menopause, including sleeping problems. Consulting with a menopause specialist is recommended.

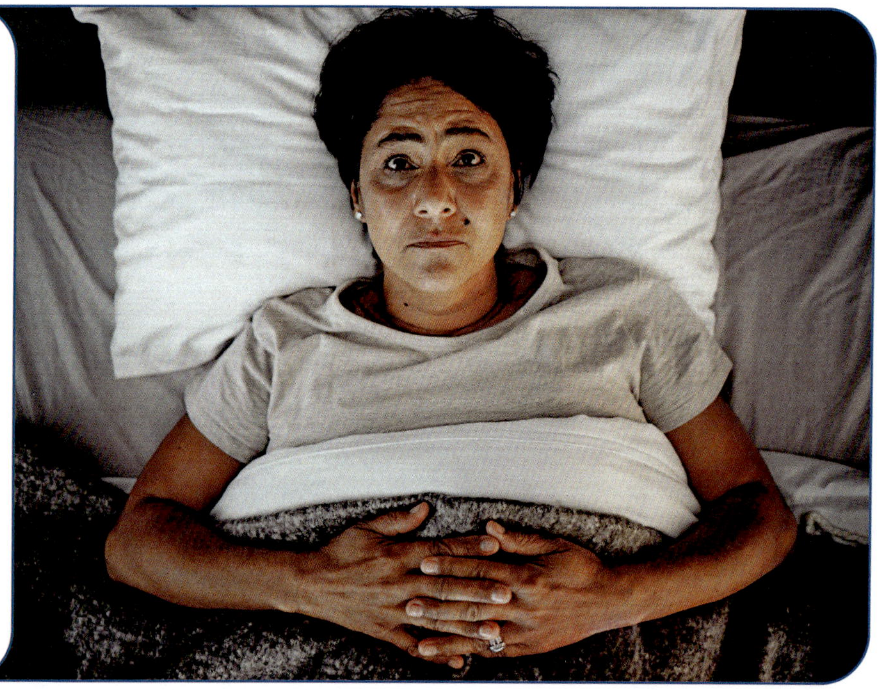

If you **opt for hysterectomy** during RRSO, you will be a candidate for estrogen-only menopausal hormone therapy. Using estrogen alone is linked with a decreased risk of breast cancer compared to the use of both estrogen and progestin.

Risk-reducing salpingectomy

Surgery to remove the fallopian tubes is called salpingectomy. If you are premenopausal and not ready to lose your ovaries, removing the fallopian tubes first and the ovaries later may be an option. But, this approach is **not** the standard of care for people at high risk of ovarian cancer.

While salpingectomy has been shown to lower ovarian cancer risk in the general population, it hasn't been proven to lower risk in those at high risk. In those at high risk, removing the ovaries and fallopian tubes is recommended. This is called a risk-reducing salpingo-oophorectomy (RRSO).

After the fallopian tubes are removed, consider continuing your combination birth control pills or hormonal IUD. This may keep your ovarian cancer risk low while you still have your ovaries.

Salpingectomy is an option for those assigned female at birth who also want tubal ligation. This form of surgical sterilization prevents pregnancy by stopping the sperm and egg from meeting.

If you freeze eggs or embryos, to become pregnant later you'll need your uterus but not your ovaries or fallopian tubes.

Pancreatic cancer

Pancreatic cancer starts in a digestive gland called the pancreas. The pancreas makes enzymes that break down proteins, fats, sugars, and starches. It also makes hormones called insulin and glucagon that control blood sugar levels.

There are 2 main types of pancreatic cancer. The more common type forms from exocrine cells. Exocrine cells make and move digestive enzymes.

A hereditary cause of pancreatic cancer is found in about 1 in 10 people. Genetic testing is recommended for everyone diagnosed with exocrine pancreatic cancer. Pancreatic cancer susceptibility genes include *ATM*, *BRCA1*, *BRCA2*, *CDKN2A*, *PALB2*, *STK11*, *TP53,* and Lynch syndrome genes (*MLH1*, *MSH2*, *MSH6*, *EPCAM*).

Hereditary pancreatitis is sometimes caused by germline variants in *PRSS1*, *SPINK1*, and related genes. While these individuals are also at increased risk of pancreatic cancer, germline testing for these genes is only recommended **if you also have** a personal or family history of exocrine pancreatic cancer.

Care for positive results is described on the next page.

If results are positive

If genetic testing finds cancer-causing variants in any susceptibility genes, your provider may recommend yearly screening for pancreatic cancer once you reach a certain age.

If pancreatic surveillance is planned, recommended methods include contrast-enhanced magnetic resonance cholangiopancreatography (MRCP) and endoscopic ultrasound (EUS).

MRCP is a type of MRI that makes very clear pictures of the pancreas and bile ducts. Contrast is used to help identify small tumors. EUS is similar to an upper endoscopy, except that the scope also has an ultrasound component. The ultrasound probe on the endoscope makes images of the pancreas and nearby structures.

If pancreatic surveillance is planned, recommendations on when to start depend on the affected gene(s). See **Guide 1.**

Guide 1
Suggested screening approach for exocrine pancreatic cancer

Germline variant in *STK11*	Consider beginning screening at age 30 to 35*
Germline variant in *CDKN2A*	Consider beginning screening at age 40*
Germline variant in *ATM* or *BRCA2*	Consider beginning screening at age 50*
Germline variant in *BRCA1, MLH1, MSH2, MSH6, EPCAM, PALB2,* or *TP53*	If you have 1 or more first- or second-degree relatives on the same side of the family with exocrine pancreatic cancer, consider beginning screening at age 50*.

**Or 10 years before the earliest exocrine pancreatic cancer diagnosis in your family, whichever is earlier*

Prostate cancer

The prostate is a gland located deep inside the pelvis. In the United States, prostate cancer is the second most common cancer in people assigned male at birth.

Prostate cancer susceptibility genes include *ATM*, *BRCA1/2*, *CHEK2*, *HOXB13*, and *TP53*.

Who should be tested?

In people diagnosed with prostate cancer, genetic testing is recommended for those who meet the criteria in **Guide 2**. If you don't have prostate cancer, or you do but don't meet these criteria, testing is recommended if you have a first-degree blood relative who *does* meet these criteria.

Who might be tested?

Testing may also be appropriate if:

- You were diagnosed with prostate cancer at or before age 55 and don't meet testing criteria
- You were diagnosed with intermediate-risk prostate cancer with intraductal/cribriform histology at any age

Guide 2
Testing criteria for people diagnosed with prostate cancer (at any age)

Ancestry	Genetic testing is recommended for people of Ashkenazi Jewish ancestry.
Cancer features	Genetic testing is recommended for people with: • metastatic prostate cancer (stage 4B) • prostate cancer that has spread to lymph nodes (stage 4A) • very high-risk or high-risk prostate cancer

Family history	Genetic testing is recommended in the following situations:
	• You have 1 or more close blood relatives with any of the following:
	  • breast cancer diagnosed at age 50 or earlier
	  • male breast cancer
	  • ovarian cancer
	  • pancreatic cancer
	  • prostate cancer that is high-risk, very high-risk, has spread to lymph nodes, or is metastatic
	• You have 3 or more close relatives (including you) with prostate cancer or breast cancer on the same side of the family.

If results are positive

Based on your specific results, your provider may recommend screening for prostate cancer by checking your prostate specific antigen (PSA) level. More info on PSA is provided below.

In the past, screening for prostate cancer involved getting a digital rectal exam (DRE). Now, DRE is generally only performed if needed.

PSA

PSA is a protein made inside the prostate gland. Its job is to help semen transport sperm. All prostate cells, both normal cells and cancer cells, make PSA.

If there's something wrong with the prostate—like prostate cancer—the prostate may make more PSA. While most PSA goes into semen, a little bit ends up in the bloodstream, too. An unusually high amount of PSA in the blood may be a sign of prostate cancer.

However, age and other factors—such as an enlarged prostate or a urinary tract infection—can also cause high levels of PSA.

We want your feedback!

Our goal is to provide helpful and easy-to-understand information on cancer. Take our survey to let us know what we got right and what we could do better. NCCN.org/patients/feedback

Li-Fraumeni syndrome

Li-Fraumeni syndrome (LFS) is caused by a pathogenic variant in the *TP53* gene. This gene helps control repair or survival of damaged cells.

Families with Li-Fraumeni syndrome have a high risk of developing a range of cancers. The 5 most common or "core" LFS cancers include:

- Sarcomas
- Leukemias
- Brain cancers
- Breast cancers
- Adrenocortical tumors

Who should have testing?

Genetic testing is recommended in the following situations:

- A germline *TP53* variant runs in your family
- You meet the classic criteria or the Chompret criteria (see **Guide 3**)
- You or someone in your family has been diagnosed with pediatric hypodiploid leukemia

Genetic testing may also be recommended if the tumor has a *TP53* mutation. Keep reading for more information.

Guide 3 Criteria for diagnosing LFS
Classic criteria
You were diagnosed with sarcoma before age 45
and
You have a first-degree relative diagnosed with cancer before age 45
and
You have a first- or second-degree relative in the same line diagnosed with cancer by age 45 or with a sarcoma at any age
Chompret criteria
You were diagnosed with a tumor in the LFS spectrum before age 46, and you have at least 1 first- or second-degree relative diagnosed with a cancer in the LFS spectrum before age 56, or with multiple primary cancers at any age
or
You've had more than 1 type of cancer and 2 of them are in the LFS spectrum. The first was found before age 46.
or
You were diagnosed with 1 of several rare cancers
or
You were diagnosed with breast cancer before age 31

Tumors with the TP53 biomarker

It's possible--and even common--for someone born without a germline *TP53* variant to have a tumor with a somatic (acquired) *TP53* mutation. For example, unlike other types of sarcoma, germline *TP53* variants are rare in those with Ewing sarcoma, gastrointestinal stromal tumors (GIST), desmoid tumors, or angiosarcoma.

If you don't meet testing criteria but the cancer has a *TP53* mutation, genetic testing might be ordered. Your personal and family history will be reviewed in order to determine if genetic testing is worthwhile.

Testing is typically ordered for those diagnosed with any cancer before age 30, or if your provider feels that it's appropriate.

Adults with positive results

Li-Fraumeni syndrome is rare. Referral to a specialized team or center with expertise in this syndrome is recommended.

People with LFS are at high risk of developing more than one type of cancer. Unfortunately, there isn't a good screening method for all of them.

See **Guide 4** on the next page for cancer screening recommendations in adults. More frequent screening may be appropriate based on your health and family history.

Whole-body MRI

Screening with whole body MRI can find cancer early in families with Li-Fraumeni syndrome. But, it can also flag things as cancer that aren't (a false-positive result).

If whole-body MRI isn't available to you, ask your provider about available clinical trials you may be able to join. Also ask about other advanced imaging methods that may be an option.

LFS and radiation therapy

Radiation increases the risk of developing a sarcoma. If you are receiving treatment for cancer, radiation therapy is typically avoided if you have other good treatment options.

But sometimes radiation therapy is needed. In this case, make an informed decision by fully discussing the benefits of radiation compared to the risk of second cancers caused by radiation.

Children with positive results

If LFS runs in your family, make your pediatrician aware of the risk of childhood cancers.

In families with LFS, the following screening measures are recommended to help detect cancer early in children:

- Complete physical exam including neurologic exam every 6 to 12 months
- Yearly whole body MRI starting in infancy
- Yearly brain MRI starting in infancy (may be performed as part of the whole body MRI or separately)
- For adrenocortical carcinoma, ultrasound every 3 to 4 months starting in infancy

Guide 4
Cancer screening in adults with germline *TP53* variants

Screening for breast cancer in those assigned female at birth	**Age 18:** Start becoming familiar with your breasts and report any changes to your health care provider. Doing breast self exams on a regular basis may help you notice changes. If you are premenopausal, doing the exams when your period is almost over may be most helpful. **Starting at age 20:** Clinical breast exam every 6 to 12 months (or, if someone in your family was diagnosed before age 20, start at that age) **Age 20 to 29:** Yearly breast MRI (or, if someone in your family was diagnosed before age 20, start at that age) **Age 30 to 75:** Yearly breast MRI and mammogram **Age 76 and above:** Care is individualized **Risk-reducing mastectomy** is an option for those with germline *TP53* variants. Your provider will explain the risks of surgery, the extent of cancer protection it provides, the possible quality-of-life effects, and your options for breast reconstruction.
Screening for other cancers	**General recommendations** • Complete physical exam including neurologic exam every 6 to 12 months for cancer survivors. Your provider will look closely for any signs or symptoms of rare and second cancers. • Yearly whole-body and brain MRI **Skin cancer** Annual skin exam starting at age 18 **Colon and stomach cancer** Colonoscopy and upper endoscopy every 2 to 5 years starting at age 25, or 5 years before the earliest colorectal or gastric cancer in the family. If your abdomen or whole body was treated with radiation therapy, colonoscopy is recommended starting 5 years after treatment. **Prostate cancer** Annual prostate-specific antigen (PSA) test starting at age 40

Cowden syndrome/PHTS

Germline variants in the *PTEN* gene can cause a spectrum of disorders called *PTEN* hamartoma tumor syndrome (PHTS). PHTS includes the following disorders:

- Cowden syndrome (CS)
- Bannayan-Riley-Ruvalcaba syndrome (BRRS)
- Adult L'hermitte-Duclos disease (LDD)
- Proteus-like syndrome
- Autism spectrum disorders with macrocephaly

PHTS syndromes cause benign (non-cancerous) tumors called hamartomas to form in or on the body. In addition to benign tumors, people with Cowden syndrome are at increased risk of malignant (cancerous) tumors or lesions, most often in the skin, mucus membranes, breasts, thyroid, endometrium, and brain.

About 1 in 2 individuals assigned female at birth with Cowden syndrome will develop breast cancer, usually by age 50.

Cowden syndrome can be diagnosed based on signs, symptoms, and health history. This is called a clinical diagnosis. A clinical diagnosis is made if you have certain combinations of the health conditions listed in **Guide 5**.

Genetic testing is recommended for people with a clinical diagnosis of CS/PHTS. Other criteria is described next.

Guide 5 **Cowden syndrome features**
Autism spectrum disorder
Breast cancer
Colon cancer
Endometrial cancer
Esophageal glycogenic acanthoses (thickened, raised areas on the inner lining of the esophagus)
Gastrointestinal hamartomas (benign tumors with a distorted structure made of an abnormal mixture of cells and tissue found in that area)
Gastrointestinal ganglioneuromas (benign nerve cell tumors)
Hyperpigmentation (darkening of the skin) on the glans penis
Intellectual disability
Kidney cancer (renal cell carcinoma)
L'hermitte-Duclos disease
Lipomas (benign fatty tumors)
Macrocephaly (enlarged head)
Thyroid cancer (papillary or follicular)
Thyroid problems (nodules, goiter)
Testicular lipomatosis (benign lesions in the testicles)
Trichilemmomas (benign tumors that form in hair follicles)
Vascular problems

Testing criteria

Testing for germline *PTEN* variants is recommended in the following situations:

- You meet the criteria for a clinical diagnosis of CS/PHTS
- A *PTEN* variant runs in your family
- You have BRRS
- You don't meet the criteria for a clinical diagnosis, but have certain combinations of other features in **Guide 5**
- Biomarker testing found a somatic *PTEN* variant in a tumor

If results are positive

If available to you, seek care from a specialized team or center with expertise in this rare syndrome.

A yearly, comprehensive physical exam is recommended. The exams should start at age 18, or 5 years before the youngest cancer diagnosis in the family (whichever comes first).

Cancer screening recommendations for people with CS/PHTS are provided in **Guides 6 and 7**.

Guide 6
Managing breast cancer risk for adults with a germline *PTEN* variant

Breast awareness	At age 18, start becoming familiar with your breasts and report any changes to your health care provider right away. Doing breast self exams on a regular basis may help you get familiar with your breasts and notice changes.
Cancer screening	• A clinical breast exam is recommended every 6 to 12 months starting at age 25, or 5 to 10 years before the earliest breast cancer diagnosis in the family (whichever comes first). • A mammogram and breast MRI are recommended every year starting at age 30, or 10 years before the earliest breast cancer diagnosis in the family (whichever comes first). • Screening and management are individualized for those above age 75.
Risk-reducing mastectomy	Risk-reducing mastectomy is an option. Your provider will explain the risks of surgery, the extent of cancer protection it provides, and your options for breast reconstruction.

Guide 7
Screening for other cancers in adults with a germline *PTEN* variant

Endometrial cancer	• Endometrial cancer can often be found early based on symptoms. Tell your care team about any symptoms, like bleeding between cycles or after menopause. Tracking changes in your menstrual cycle on a calendar can be helpful. • In people with CS/PHTS, screening for endometrial cancer isn't generally recommended. But, some providers suggest an endometrial biopsy every 1 to 2 years, starting by age 35. • Talk to your provider about the option of hysterectomy after childbearing. • People with CS/PHTS alone are at average risk of ovarian cancer, so there's no need to remove the ovaries.
Colorectal cancer	• Colonoscopy is recommended starting at age 35 if you don't have any signs or symptoms. • If a close relative was diagnosed with colorectal cancer before age 40, start screening 5 to 10 years before the earliest CRC diagnosis in the family. • Colonoscopy should be done every 5 years, or more frequently if patient is symptomatic or polyps are found.
Kidney cancer	Your provider may recommend a kidney ultrasound every 1 to 2 years, starting at age 40.
Skin cancer	There may be an increased risk of melanoma and other skin problems. Annual dermatology exams are recommended.
Thyroid cancer	A yearly thyroid ultrasound is recommended starting at age 7 in most cases.

Gene-specific care

The charts that follow provide recommendations for lowering your cancer risk based on your genetic testing results.

The recommendations are presented alphabetically by gene.

Guide 8
Recommendations for managing cancer risk, according to gene

ATM

Increased risk of	Recommended steps to lower your risk
Breast cancer	Get a yearly mammogram starting at age 40. Your provider may recommend breast MRI starting at age 30 or 35, based on your age, family history, breast density, and preferences.
Epithelial ovarian cancer	There isn't enough evidence to recommend risk-reducing salpingo-oophorectomy (RRSO). Care is guided by your family history.
Pancreatic cancer	See page 25 for information on pancreatic cancer screening.
Prostate cancer	There is growing evidence of a link between the *ATM* gene and prostate cancer. Your provider may check your PSA level starting at age 40.
Other cancer risks	People with this variant are at increased risk of colorectal cancer. Speak with your provider about screening recommendations and other care.

BARD1

Increased risk of	Recommended steps to lower your risk
Breast cancer	Get a yearly mammogram starting at age 40. Your provider may recommend breast MRI (in addition to mammography), also starting at age 40, based on your age, family history, breast density, and preferences.

BRCA1 and *BRCA2*

Increased risk of	Recommended steps to lower your risk
Breast cancer (in those assigned female at birth)	• Age 18: Start becoming familiar with your breasts. Self exams can help. • Age 25: Start having clinical breast exams every 6 to 12 months. • Age 25 to 29: Get a yearly breast MRI. Or, if a family member was diagnosed with breast cancer before age 30, care is personalized. • Age 30 to 75: Get a yearly mammogram **and** breast MRI • Age 75 and after: Your care is individualized • Risk-reducing mastectomy and risk-reducing agents are options.
Breast cancer (in those assigned male at birth)	• Breast self-exam training and education starting at age 35 years • Clinical breast exam every 12 months, starting at age 35 years • Your provider may recommend a yearly mammogram, starting at age 50 or 10 years before the earliest male breast cancer in the family (whichever comes first).
Ovarian cancer	• If desired, speak with a fertility specialist to discuss family planning • Nonsurgical options to lower risk include birth control pills (with both estrogen and progestin) and levonorgestrel IUDs • For *BRCA1* carriers, RRSO is recommended between age 35 and 40. For *BRCA2* carriers, it may be delayed until age 40 to 45. • Option of hysterectomy at time of RRSO • Menopausal hormone therapy can help with menopause symptoms • Removing the fallopian tubes first (and the ovaries later) may be an option for those not ready to lose their ovaries and who want tubal ligation.
Melanoma	There are no specific screening recommendations. Limiting UV exposure and getting a yearly full-body skin exam can help lower your risk.
Pancreatic cancer	See page 25 for information on pancreatic cancer screening.
Prostate cancer	For *BRCA2* carriers, screening is recommended starting at age 40. Screening **may be** recommended for *BRCA1* carriers (also at age 40).

BRIP1	
Increased risk of	**Recommended steps to lower your risk**
Epithelial ovarian cancer	Risk-reducing salpingo-oophorectomy is recommended around age 45 to 50.

CDH1	
Increased risk of	**Recommended steps to lower your risk**
Breast cancer	• Get a yearly mammogram starting at age 30. • Your provider may recommend breast MRI in addition to mammograms. • Risk-reducing mastectomy is an option.
Other cancer risks	People with this variant are at **increased risk** of hereditary diffuse gastric cancer (HDGC). Ask your provider what this means for your care.

CDKN2A	
Increased risk of	**Recommended steps to lower your risk**
Melanoma	A whole-body skin exam by a dermatologist, including total body photography and dermoscopy, is recommended every 6 months.
Pancreatic cancer	See page 25 for information on pancreatic cancer screening.
Other cancer risks	Nerve sheath tumors, sarcomas, and other cancers are also possible. Your provider may suggest surveillance beyond screening for pancreatic cancer and melanoma. This might include yearly full-body and brain MRI.

CHEK2

Increased risk of	Recommended steps to lower your risk
Breast cancer	• Get a yearly mammogram starting at age 40 • Considering your age, family history, breast density, and preferences, your provider may recommend breast MRI starting at age 30 to 35.
Prostate cancer	There is growing evidence of a link between the *CHEK2* gene and prostate cancer. Your provider may recommend checking your PSA starting at age 40.

MSH2, MLH1, MSH6, PMS2, EPCAM (associated with Lynch syndrome)

Lynch syndrome is the most common cause of hereditary colon and endometrial (uterine) cancers. Families with Lynch syndrome are also at increased risk of urothelial, brain, gastrointestinal, ovarian, breast, pancreatic, biliary, and other cancers.

The criteria for testing for Lynch syndrome and care for a positive result is beyond the scope of this guide. For more information, health care providers are encouraged to refer to the *NCCN Clinical Practice Guidelines in Oncology for Genetic/Familial High-Risk Assessment: Colorectal, Endometrial, and Gastric*.

NF1

Increased risk of	Recommended steps to lower your risk
Breast cancer	• Get a yearly mammogram starting at age 30 • Your provider may also recommend breast MRI from age 30 to 50
Other cancer risks	You may be at increased risk of malignant peripheral nerve sheath tumors, gastrointestinal stromal tumors (GIST), and others. Seeing an *NF1* specialist is recommended.

PALB2

Increased risk of	Recommended steps to lower your risk
Breast cancer	• Get a yearly mammogram **and** breast MRI starting at age 30 • Discuss the option of risk-reducing mastectomy with your provider. • For those assigned male at birth, your provider may recommend the same breast cancer screening strategy recommended for *BRCA* carriers.
Epithelial ovarian cancer	Talk to your provider about the option of risk-reducing salpingo-oophorectomy around age 45 or 50.
Pancreatic cancer	See page 25 for information on pancreatic cancer screening.

PTEN (associated with Cowden syndrome)

People with germline *PTEN* variants are at increased risk of breast, thyroid, colorectal, endometrial, and kidney cancer. See page 32 for information on Cowden syndrome/PTEN hamartoma tumor syndrome (PHTS).

RAD51C and *RAD51D*

Increased risk of	Recommended steps to lower your risk
Breast cancer	Get a yearly mammogram (and possibly also breast MRI) starting at age 40. Other care is guided by your family history.
Epithelial ovarian cancer	Risk-reducing salpingo-oophorectomy is recommended around age 45 or 50.

STK11 (associated with Peutz-Jeghers syndrome)

Peutz-Jeghers syndrome (PJS) is a hereditary cancer syndrome that causes benign tumors to grow in the gastrointestinal tract and dark spots to form on the skin. People with PJS have a high risk of developing cancers of the **breast, colon, stomach, ovaries, small intestine, pancreas, cervix, uterus, lungs, and testes**. Most people with PJS have a cancer-causing germline variant in the *STK11* gene.

Genetic testing for PJS and care for a positive result are beyond the scope of this guide. For more information, health care providers are encouraged to refer to the *NCCN Clinical Practice Guidelines in Oncology for Genetic/Familial High-Risk Assessment: Colorectal, Endometrial, and Gastric*. See below for information on managing breast and pancreatic cancer risk in people with this variant.

Breast cancer	• Get a yearly mammogram **and** breast MRI starting at age 30. • Discuss the option of risk-reducing mastectomy with your provider.
Pancreatic cancer	See page 25 for information on pancreatic cancer screening.

TP53 (associated with Li-Fraumeni syndrome)

Li-Fraumeni syndrome (LFS) is caused by a pathogenic variant in the *TP53* gene. Families with LFS have a high risk of developing a range of cancers, especially **sarcomas, leukemias, brain cancers, breast cancers, and adrenocortical tumors**.

Many other cancers have been linked with LFS, including melanoma, colorectal, gastric, and prostate cancers. See page 29 for information on Li-Fraumeni syndrome.

Key points

- In people diagnosed with breast cancer, genetic testing decisions are based on age, ancestry, family history, cancer type, and cancer features. If results are positive, your provider may recommend managing your risk with clinical breast exams, mammograms, breast MRIs, risk-reducing mastectomy, or risk-reducing drugs.
- Genetic testing is recommended for everyone diagnosed with epithelial ovarian cancer (at any age), and for their first- and second-degree blood relatives. If results are positive, risk-reducing salpingo-oophorectomy may be recommended to lower your risk. Menopausal hormone therapy (MHT) may be an option to relieve symptoms of surgical menopause.
- Genetic testing is recommended for everyone diagnosed with exocrine pancreatic cancer. If results are positive, your provider may recommend yearly screening with endoscopic ultrasound or magnetic resonance cholangiopancreatography once you reach a certain age.
- In people diagnosed with prostate cancer, genetic testing decisions are based on ancestry, family history, cancer stage, and cancer risk level. If results are positive, your provider may recommend blood tests to check your prostate specific antigen (PSA) level.
- Genetic testing for Li-Fraumeni syndrome is recommended if you meet specific criteria, if a germline *TP53* variant runs in your family, and in some other situations. In adults with a positive result, frequent physical exams and yearly MRIs (whole-body and brain) are recommended. Mammograms and breast MRIs are recommended to manage breast cancer risk. Risk-reducing mastectomy is also an option.
- Genetic testing is recommended for those with a clinical diagnosis of Cowden syndrome/*PTEN* hamartoma tumor syndrome and in some other situations. If results are positive, care to manage your risk involves yearly physical exams, staying alert for signs of endometrial cancer, and screening for breast, colorectal, kidney, skin, and thyroid cancers. Risk-reducing mastectomy is an option to lower breast cancer risk.

Questions to ask

- Based on my results, which cancers am I at risk for?
- Does having risk-reducing surgery to prevent ovarian or breast cancer mean that I'll never have to worry about getting those cancers?
- Is risk-reducing mastectomy something I need to consider? If so, who can help me decide?

4

Other resources

Want to learn more? Here's how you can get additional help.

What else to know

This book can help you improve your cancer care. It plainly explains expert recommendations and suggests questions to ask your care team. But, it's not the only resource that you have.

You're welcome to receive as much information and help as you need. Many people are interested in learning more about:

- The details of their health and treatment
- Being a part of a care team
- Getting financial help
- Finding a care provider who is an expert in their field
- Coping with side effects

What else to do

Your health care center can help you with next steps. They often have on-site resources to help meet your needs and find answers to your questions. Health care centers can also inform you of resources in your community.

In addition to help from your providers, the resources listed in the next section provide support for many people like yourself. Look through the list and visit the provided websites to learn more about these organizations.

Where to get help

Breastcancer.org
Breastcancer.org

Breast Cancer Alliance
Breastcanceralliance.org

CanCare, Inc.
Cancare.org

Cancer*Care*
Cancercare.org

Cancer Hope Network
cancerhopenetwork.org

DiepC Foundation
diepcfoundation.org

FORCE: Facing Our Risk of Cancer Empowered
Facingourrisk.org

GPAC Global Patient Advocacy Coalition
GPACunited.org

HIS Breast Cancer Awareness
hisbreastcancer.org

Imerman Angels
Imermanangels.org

Lustgarten Foundation
lustgarten.org

National Coalition for Cancer Survivorship
Canceradvocacy.org

National Ovarian Cancer Coalition
Ovarian.org

Ovarcome
Overcome.org

Ovarian Cancer Research Alliance
ocrahope.org

Sharsheret
sharsheret.org

Triage Cancer
Triagecancer.org

Words to know

BRCA1 and _BRCA2_ genes
Variants in these DNA repair genes put you at high risk of breast and ovarian cancer, and increased risk of prostate and pancreatic cancer.

cancer predisposition genes
Genes that—if altered in specific ways—increase your risk of developing certain types of cancer. Also called cancer susceptibility genes.

clinical breast exam (CBE)
A physical exam of the breast performed by a health care provider to check for lumps or other changes.

close blood relatives
Close blood relatives include first-, second-, and third-degree relatives on the same side of the family.

epithelial ovarian cancer
Cancer that starts in the surface layer of tissue surrounding the fallopian tubes and ovaries. The most common type of ovarian cancer.

fallopian tube
A thin tube through which an egg travels from the ovary to the uterus.

founder variant
A disease-causing variant that occurs frequently in distinct groups of people whose shared ancestor(s) carried the variant.

genetic counseling
A discussion with a health expert about the risk for a disease caused by changes in genes.

genetic testing
Testing of the blood, saliva, or other samples for germline (inherited) variants that cause cancer.

germline variant
A gene difference that is passed from a parent to their biological child(ren).

Li-Fraumeni syndrome (LFS)
A hereditary cancer syndrome caused by a cancer-causing variant in the *TP53* gene. Families with LFS have a high risk of developing sarcomas, leukemias, brain cancers, breast cancers, and adrenocortical tumors.

Lynch syndrome
A hereditary cancer syndrome that increases the risk of developing colorectal, endometrial, ovarian, and other cancers.

mammogram
A picture of the insides of the breast that is made using x-rays.

menopausal hormone therapy (MHT)
The use of hormones to lessen side effects of menopause. This approach used to be called hormone replacement therapy (HRT).

next-generation sequencing (NGS)
Testing a sample for many gene variants at one time. Also called a multi-gene panel.

pathogenic/likely pathogenic (P/LP) variant
A variant (difference) in a gene that means you are at increased risk of developing 1 or more cancers.

penetrance
Describes how likely someone with an inherited cancer-causing variant is to develop signs and symptoms of the cancer. Not everyone who has the variant will develop the cancer.

Peutz-Jeghers syndrome (PJS)
A hereditary cancer syndrome in which many polyps form in the lining of the gastrointestinal tract and dark spots appear on the skin. People with PJS have a high risk of developing cancers of the GI tract, breast, pancreas, ovary, lung, and cervix. PJS is usually caused by differences in the *STK11* gene.

risk-reducing mastectomy (RRM)
Surgery that removes the whole breast in order to lower the risk of developing breast cancer.

risk-reducing salpingo-oophorectomy (RRSO)
Surgery that removes the ovaries and fallopian tubes in order to lower the risk of ovarian cancer, fallopian tube cancer, and primary peritoneal cancers in those at high risk.

screening mammogram
X-rays of the breasts taken to check for breast cancer in someone without signs or symptoms of cancer.

somatic mutation
A non-hereditary change in your DNA that occurred sometime after you were conceived. Also called acquired mutation or tumor mutation.

surgical menopause
The forced start of menopause caused by surgery to remove the ovaries. Results from a sudden drop in estrogen in the body.

variant
An difference in a gene that may be helpful, harmful, or have no (known) effect on your health. Variants you are born with are called germline variants.

NCCN Contributors

This patient guide is based on the NCCN Clinical Practice Guidelines in Oncology (NCCN Guidelines®) for Genetic/Familial High-Risk Assessment: Breast, Ovarian, Pancreatic, and Prostate, Version 2.2025 – November 7, 2024. It was adapted, reviewed, and published with help from the following people:

Dorothy A. Shead, MS
Senior Director
Patient Information Operations

Erin Vidic, MA
Senior Medical Writer, Patient Information

Susan Kidney
Senior Graphic Design Specialist

The NCCN Guidelines® for Genetic/Familial High-Risk Assessment: Breast, Ovarian, Pancreatic, and Prostate, Version 2.2025 were developed by the following NCCN Panel Members:

*Mary B. Daly, MD, PhD/Chair
Fox Chase Cancer Center

*Tuya Pal, MD/Vice-Chair
Vanderbilt-Ingram Cancer Center

Zahraa AlHilli, MD
Case Comprehensive Cancer Center/ University Hospitals Seidman Cancer Center and Cleveland Clinic Taussig Cancer Institute

Banu Arun, MD
The University of Texas MD Anderson Cancer Center

Saundra S. Buys, MD
Huntsman Cancer Institute at the University of Utah

Heather H. Cheng, MD, PhD
Fred Hutchinson Cancer Center

*Jane Churpek, MD, MS
University of Wisconsin Carbone Cancer Center

*Sarah Colonna, MD, MSCI
Huntsman Cancer Institute at the University of Utah

Susan M. Domchek, MD
Abramson Cancer Center at the University of Pennsylvania

*Susan Friedman, DVM
FORCE: Facing Our Risk of Cancer Empowered

Veda N. Giri, MD
Yale Cancer Center/Smilow Cancer Hospital

*Michael Goggins, MD
Johns Hopkins Kimmel Cancer Center

Andrea Hagemann, MD, MSCI
Siteman Cancer Center at Barnes-Jewish Hospital and Washington University School of Medicine

*Ashley Hendrix, MD, MBA
The University of Tennessee Health Science Center

Dezheng Huo, PhD
The UChicago Medicine Comprehensive Cancer Center

Mollie L. Hutton, MS, CGC
Roswell Park Comprehensive Cancer Center

Beth Y. Karlan, MD
UCLA Jonsson Comprehensive Cancer Center

Nawal Kassem, MD, MS
Indiana University Melvin and Bren Simon Comprehensive Cancer Center

Seema Khan, MD
Robert H. Lurie Comprehensive Cancer Center of Northwestern University

Katia Khoury, MD
O'Neal Comprehensive Cancer Center at UAB

*Allison W. Kurian, MD, MSc
Stanford Cancer Institute

Christine Laronga, MD
Moffitt Cancer Center

Julie S. Mak, MS, MSc, CGC
UCSF Helen Diller Family Comprehensive Cancer Center

John Mansour, MD
UT Southwestern Simmons Comprehensive Cancer Center

Kara N. Maxwell, MD, PhD
Abramson Cancer Center at the University of Pennsylvania

Kevin McDonnell, MD, PhD
City of Hope National Medical Center

Carolyn S. Menendez, MD
Duke Cancer Institute

Sofia D. Merajver, MD, PhD
University of Michigan Rogel Cancer Center

Barbara S. Norquist, MD
Fred Hutchinson Cancer Center

Kenneth Offit, MD, MPH
Memorial Sloan Kettering Cancer Center

Dominique Rash, MD
UC San Diego Moores Cancer Center

Gwen Reiser, MS, CGC
Fred & Pamela Buffett Cancer Center

*Leigha Senter-Jamieson, MS, CGC
The Ohio State University Comprehensive Cancer Center – James Cancer Hospital and Solove Research Institute

Kristen Mahoney Shannon, MS, CGC
Mass General Cancer Center

Kala Visvanathan, MD, MHS
Johns Hopkins Kimmel Cancer Center

Jeanna Welborn, MD
UC Davis Comprehensive Cancer Center

*Myra J. Wick, MD, PhD
Mayo Clinic Comprehensive Cancer Center

Marie Wood, MD
University of Colorado Cancer Center

Matthew B. Yurgelun, MD
Dana-Farber/Brigham and Women's Cancer Center

NCCN

Susan Darlow, PhD
Senior Manager, Guidelines Information Standardization

Zeenat Diwan, MS, PhD
Associate Scientist/Medical Writer

Mary Dwyer, MS
Senior Director, Guidelines Operations

* Reviewed this patient guide. For disclosures, visit NCCN.org/disclosures.

NCCN Cancer Centers

Abramson Cancer Center
at the University of Pennsylvania
Philadelphia, Pennsylvania
800.789.7366 • pennmedicine.org/cancer

Case Comprehensive Cancer Center/
University Hospitals Seidman Cancer Center and
Cleveland Clinic Taussig Cancer Institute
Cleveland, Ohio
UH Seidman Cancer Center
800.641.2422 • uhhospitals.org/services/cancer-services
CC Taussig Cancer Institute
866.223.8100 • my.clevelandclinic.org/departments/cancer
Case CCC
216.844.8797 • case.edu/cancer

City of Hope National Medical Center
Duarte, California
800.826.4673 • cityofhope.org

Dana-Farber/Brigham and Women's Cancer Center |
Mass General Cancer Center
Boston, Massachusetts
877.442.3324 • youhaveus.org
617.726.5130 • massgeneral.org/cancer-center

Duke Cancer Institute
Durham, North Carolina
888.275.3853 • dukecancerinstitute.org

Fox Chase Cancer Center
Philadelphia, Pennsylvania
888.369.2427 • foxchase.org

Fred & Pamela Buffett Cancer Center
Omaha, Nebraska
402.559.5600 • unmc.edu/cancercenter

Fred Hutchinson Cancer Center
Seattle, Washington
206.667.5000 • fredhutch.org

Huntsman Cancer Institute at the University of Utah
Salt Lake City, Utah
800.824.2073 • healthcare.utah.edu/huntsmancancerinstitute

Indiana University Melvin and Bren Simon
Comprehensive Cancer Center
Indianapolis, Indiana
888.600.4822 • www.cancer.iu.edu

Johns Hopkins Kimmel Cancer Center
Baltimore, Maryland
410.955.8964
www.hopkinskimmelcancercenter.org

Mayo Clinic Comprehensive Cancer Center
Phoenix/Scottsdale, Arizona
Jacksonville, Florida
Rochester, Minnesota
480.301.8000 • Arizona
904.953.0853 • Florida
507.538.3270 • Minnesota
mayoclinic.org/cancercenter

Memorial Sloan Kettering Cancer Center
New York, New York
800.525.2225 • mskcc.org

Moffitt Cancer Center
Tampa, Florida
888.663.3488 • moffitt.org

O'Neal Comprehensive Cancer Center at UAB
Birmingham, Alabama
800.822.0933 • uab.edu/onealcancercenter

Robert H. Lurie Comprehensive Cancer
Center of Northwestern University
Chicago, Illinois
866.587.4322 • cancer.northwestern.edu

Roswell Park Comprehensive Cancer Center
Buffalo, New York
877.275.7724 • roswellpark.org

Siteman Cancer Center at Barnes-Jewish Hospital
and Washington University School of Medicine
St. Louis, Missouri
800.600.3606 • siteman.wustl.edu

St. Jude Children's Research Hospital/
The University of Tennessee Health Science Center
Memphis, Tennessee
866.278.5833 • stjude.org
901.448.5500 • uthsc.edu

Stanford Cancer Institute
Stanford, California
877.668.7535 • cancer.stanford.edu

The Ohio State University Comprehensive Cancer Center -
James Cancer Hospital and Solove Research Institute
Columbus, Ohio
800.293.5066 • cancer.osu.edu

The UChicago Medicine Comprehensive Cancer Center
Chicago, Illinois
773.702.1000 • uchicagomedicine.org/cancer

The University of Texas MD Anderson Cancer Center
Houston, Texas
844.269.5922 • mdanderson.org

UC Davis Comprehensive Cancer Center
Sacramento, California
916.734.5959 • 800.770.9261
health.ucdavis.edu/cancer

UC San Diego Moores Cancer Center
La Jolla, California
858.822.6100 • cancer.ucsd.edu

UCLA Jonsson Comprehensive Cancer Center
Los Angeles, California
310.825.5268 • uclahealth.org/cancer

UCSF Helen Diller Family Comprehensive Cancer Center
San Francisco, California
800.689.8273 • cancer.ucsf.edu

University of Colorado Cancer Center
Aurora, Colorado
720.848.0300 • coloradocancercenter.org

University of Michigan Rogel Cancer Center
Ann Arbor, Michigan
800.865.1125 • rogelcancercenter.org

University of Wisconsin Carbone Cancer Center
Madison, Wisconsin
608.265.1700 • uwhealth.org/cancer

UT Southwestern Simmons Comprehensive Cancer Center
Dallas, Texas
214.648.3111 • utsouthwestern.edu/simmons

Vanderbilt-Ingram Cancer Center
Nashville, Tennessee
877.936.8422 • vicc.org

Yale Cancer Center/Smilow Cancer Hospital
New Haven, Connecticut
855.4.SMILOW • yalecancercenter.org

share with us.

Take our survey and help make the NCCN Guidelines for Patients better for everyone!

NCCN.org/patients/comments

Notes

Index

STEMD^2 Research & Development Group

Center on Disability Studies

College of Education

University of Hawai'i at Mānoa

http://stemd2.com/

Printed in the United States of America. First published (2017) by the Center on Disability Studies, University of Hawai'i at Mānoa, Honolulu, Hawai'i. This document is based upon work supported by the Department of Education, Native Hawaiian Education Act Program under award #S362A140018. Any opinions, findings, and conclusions or recommendations expressed in this material are those of the authors and do not necessarily reflect the views of the United States Department of Education. For further information about this document and Ne'epapa Ka Hana project, please contact Dr. Kaveh Abhari at abhari@hawaii.edu.

ISBN: 978-0-9983142-5-9

Second release, 2020

Ne‘epapa Ka Hana Eighth-Grade Mathematics Resources

Let’s Go Fishing

Student Activities

Project Directors	Kaveh Abhari Kelly Roberts
Content Developers	Justin Toyofuku Robert G. Young
Publication Designers	Katie Gao Robert G. Young MyLan Tran
Content Contributors	Remy Pages Kimble McCann

Acknowledgments

We would like to thank Ruth Silberstein, Cody Kikuta, Katy Parsons, Crystal Yoo, and Kelli Ching for advising on eighth-grade mathematics education.

Suggested Citation

Let’s Go Fishing (2020). Ne‘epapa Ka Hana Eighth-grade Mathematics Resources. STEMD² Book Series. STEMD² Research and Development Group. Center on Disability Studies, University of Hawai‘i at Mānoa. Honolulu, Hawai‘i.

Contents

Preface

About the STEMD² Book Series

The STEMD² Book Series for eighth-grade mathematics was developed as part of a technology-enabled pedagogical approach (Ne'epapa Ka Hana model) for teaching mathematics in Hawai'i middle schools. This book series seeks to provide Hawai'i middle school teachers resources and training to incorporate problem-based learning, social learning and inclusive pedagogy through a culturally relevant mathematics curriculum. The series currently includes:

Let's Build a Canoe – Student Activities and Teacher's Guide (Common Core aligned)
Let's Play the 'Ukulele – Student Activities and Teacher's Guide (Common Core aligned)
Let's Go Fishing – Student Activities and Teacher's Guide (SBAC aligned)
Let's Make Da Kine / E Hana Kākou – Student Mini Projects and Teacher's Guide in English and 'Ōlelo Hawai'i (Skill development)

The printed and online resources produced by NKH through STEMD² are fully aligned with the Common Core Standards for Mathematical Practice and Content and the Smarter Balanced Assessments for mathematics. Based on the on GO Math!® curriculum structure, the NKH STEMD² book series and social learning platform (`community.stemd2.com`) is flexible for teachers to implement partially or fully in their classrooms, as a tool to encourage students' interest and achievement in STEM subjects.

Funded by a three-year grant from the Department of Education's Native Hawaiian Education Act Program, Ne'epapa Ka Hana is a project of the STEMD² Research & Development Group in the Center on Disability Studies at the University of Hawai'i at Mānoa. More information about STEMD² is available online at `www.stemd2.com`.

Unit 1: The Number System

Activity 1.1

Mauna Kea is about 1.38×10^4 feet tall. Yellowfin tuna, or better known as ahi in the islands, have been recorded to reach depths of 3.8×10^3 feet in the ocean. How much higher is Mauna Kea than the recorded depth an ahi can reach?

Activity 1.2

You and your friends have kept track of all the hours each of you went fishing for the past 5 years (Y), and how many fish each of you caught (X). Find out who has the highest ratio of *number of fish caught* to *number of hours spent fishing* by filling out the table and comparing their ratios. Keep in mind there are different types of fishing, some people scoop nehu (small bait fish) and count that too.

	Number of fish caught (X)	Number of hours spent fishing (Y)	Ratio $\left(\frac{X}{Y}\right)$
You	3^4	3^5	
Robert	2^2	2^{-2}	
Sara	2	$\left(\frac{1}{2}\right)^2$	
Mike	$(-5)^4$	$(-5)^6$	

Who had the highest ratio?

Activity 1.3

The area of the island of Oahu can be estimated by looking at its greatest dimensions, which is about 7×10^7 millimeters long and about 5×10^7 millimeters wide. If you wanted to make a square net with *enough area to cover the entire island*, how long should each side be? So if d^2 is the area of the island of Oahu, solve the following to determine the length d. Please round your answer to the nearest million millimeters, then write it in scientific notation.
Hint: Area of net $= d^2 = (7 \times 10^7)(5 \times 10^7)$.

Activity 1.4

The price for fresh poke on the island varies with the supply of fish available. Some of these fish are bought fresh from local fishermen. Unfortunately, today the market's scale was broken and can only weigh things in grams. The worker informs you that today's market price for ahi is $17.64 per kg ($1 \times 10^3$g $= 1$kg).

The following table shows a great day's catch (4 ahi) and how much you're offered to sell your fish to the market. Wanting to become a more experienced fisherman who sells fish to the market, you would like to calculate the ahi's weight for future reference. Does each fish weigh between 4.219×10^4 and 53.21×10^3 grams?

Fish	Price of fish ($)	Weight of fish (grams)	Between 4.219×10^4 and 53.21×10^3? (Write *Yes* or *No*)
#1	800.00		
#2	920.20		
#3	760.00		
#4	640.00		

Unit 2: Proportional and Nonproportional Relationships and Functions

Activity 2.1

It is usually much safer to go fishing with your friends than by yourself. A lot of dangerous things can happen while out on the ocean, so it's good to have some friends around who can help. For example, a common rule for spearfishing is to have one diver underwater while their partner stays on the surface, waiting to help if needed. Fill out the two tables below. Both tables should show that more lives are saved when there are more people present. The first table should show a function between *people present* and *lives saved*, but the second table should not be a function.

Function

People present (number)	Lives saved (number)

VS.

Not a Function

People present (number)	Lives saved (number)

Activity 2.2

There's a graduation party coming up at the end of the summer, and you want to provide fish for your friends and family who attend. So you decide to go fishing for *oama* (juvenile goatfish) the week before the party. You plan to catch the bag limit of 50 oama per day to get to your goal of 200 oama for the party.

Here's a recap of how each fishing day went:

1. The oama weren't biting for half the day, but then the pace of your catch picked up nicely and consistently.

2. As soon as you got to your fishing spot, a nice fishermen gave you 50 oama because he had caught too many. Instead of taking it all home, you decided to use some for bait and fish for fun. At the end of the day, you got back to fishing for oama to replace the ones that you've used.

3. The oama was biting consistently throughout the day and you caught your limit for the day.

4. The oama were biting well and you caught 30 oama quickly, but then you gave some away to another fishermen. Later, you finished catching your limit.

Each of the following graphs represent the amount of oama you need catch for the day (y) vs time (x). Match each graph with a story from above (1-4).

Graph A

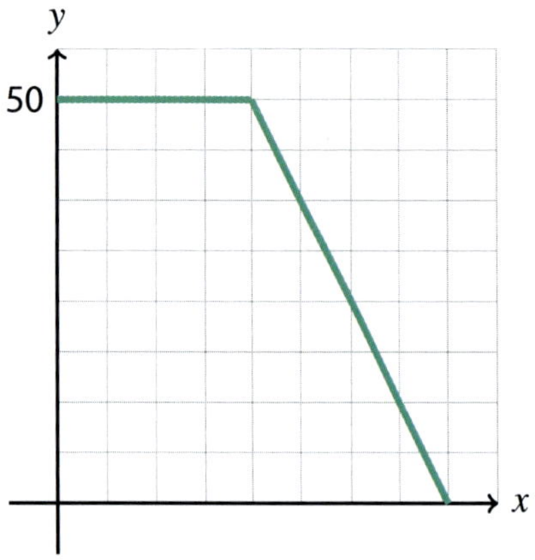

Graph B

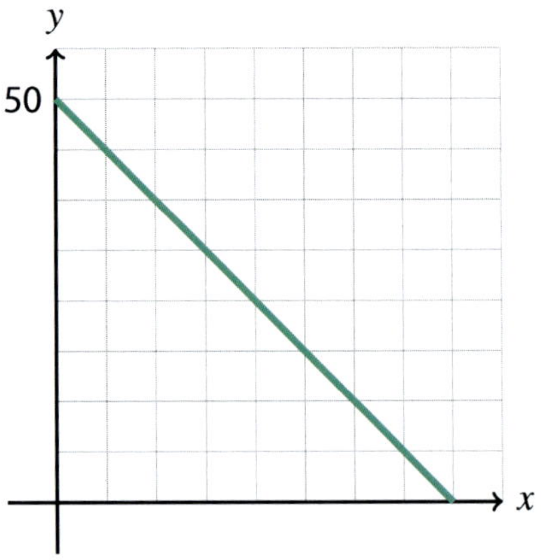

Graph C

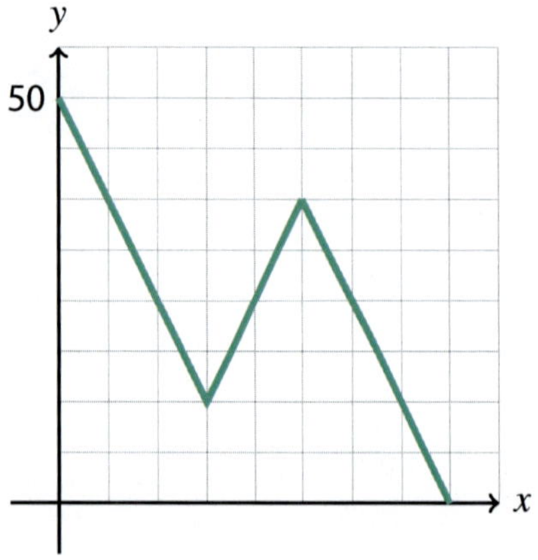

Graph D

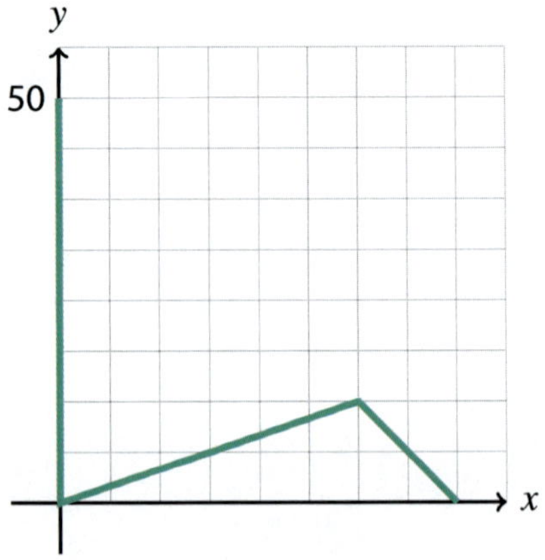

Activity 2.3

Finding fish can be hard at times, but there are places where fish tend to hang out. One of these places is along ledges and steep drop offs of the reef. Below is a bird's eye view of a reef with a line showing its drop off. The circle at (2, 4) is a good place to catch fish. What are three other coordinates that are good places to find fish?

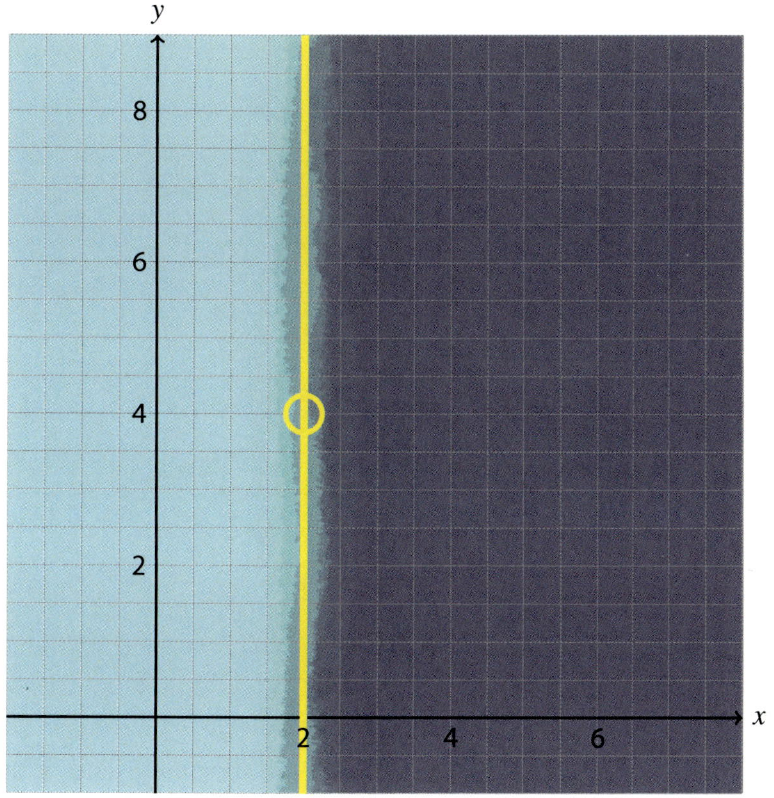

Activity 2.4

A modern GPS (Global Positioning System) can accurately pinpoint your location and track your progress throughout a day. Below is the GPS data from your last two fishing trips.

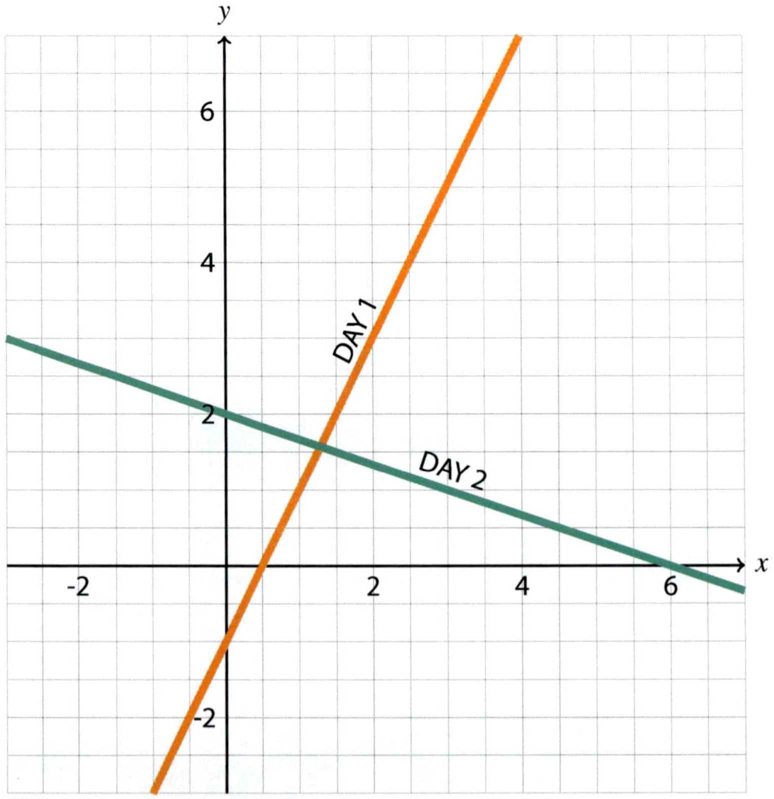

Find the equations to best describe your path for the Day 1 (the orange line) and Day 2 (the blue line) to let your friends know where the fish was biting.

Are these equations examples of linear functions?

Activity 2.5

A *live well* is a container with a water pump used for keeping fish alive, typically fish that will be used as bait. Suppose that your live well is cylindrical with a height of 10 inches and a radius of 4 inches. Your water pump can bring in new salt water at a rate of 100 cubic inches per minute. At this rate, how long will it take you to completely fill your live well without it overflowing? Round your answer to the nearest minute.

Activity 2.6

Oama season has come again this summer, and here is how the day went fishing for you:

- For the first hour, the fish were biting fast, then it slowed down, You caught 10 oama.
- Over the next two hours, the fish were biting at a steady rate until you had 25 oama total in your cooler.
- Unfortunately, for the next hour, the fish weren't biting at all.
- In the hour after that, the fish started biting slowly. Luckily, the fishing picked up, and you now 40 in the cooler.
- Over your final hour fishing, you steadily caught your last 10 oama, reaching your bag limit for the day.

On the Cartesian plane below, graph how this day of fishing might look in terms of the number of fish caught over time.

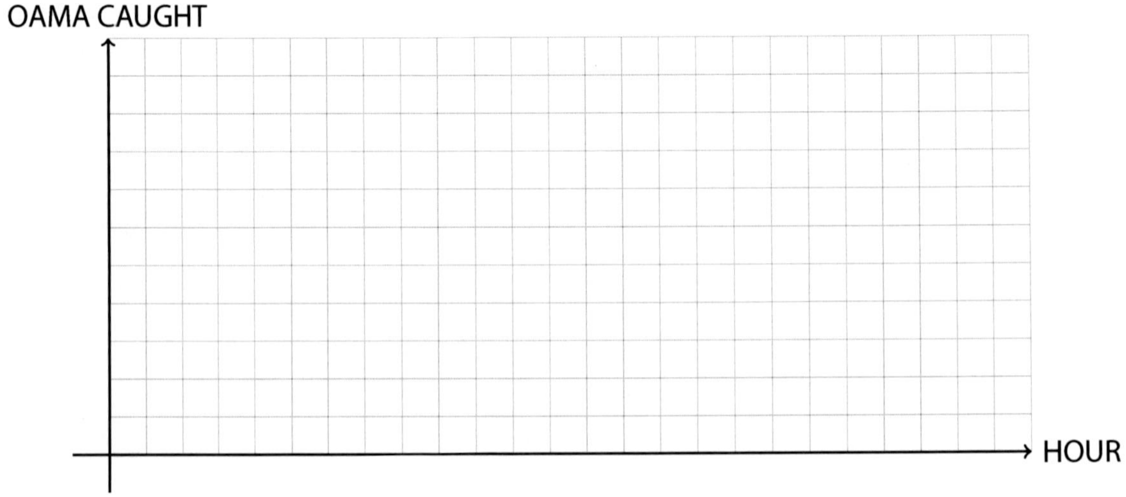

Activity 2.7

You and two of your friends are saving up for some new fishing gear. Kainoa saved up his money and showed you how much he saved on a graph.

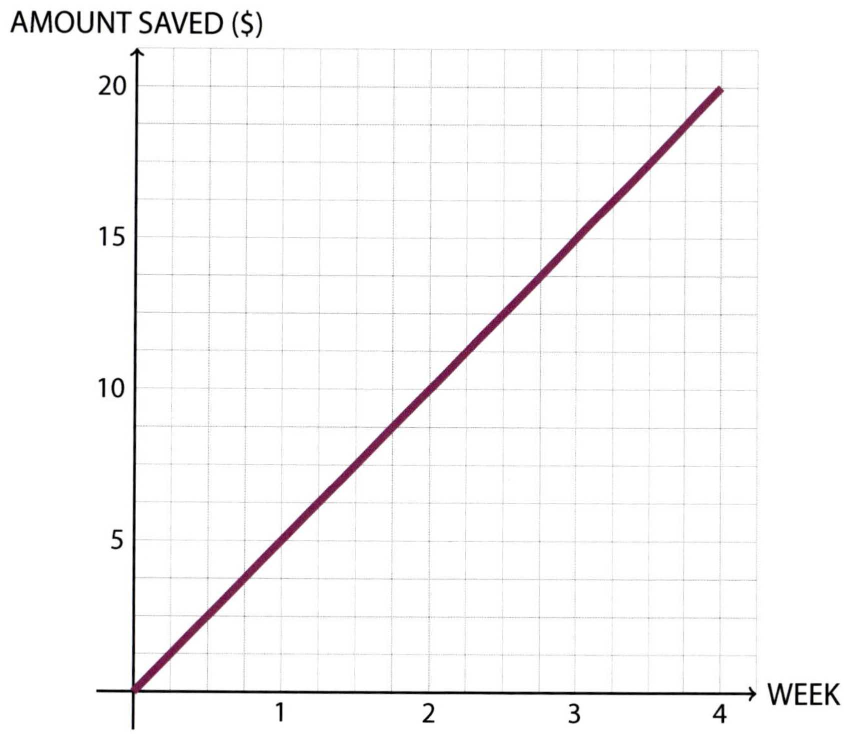

Lilo saved up some money too, but showed you her savings in a table:

Week	Total amount saved
1	$4.50
2	$9.00
3	$13.50
4	$18.00

You also saved some money, but you saved the same amount of money every week for 4 weeks. You modeled your savings with the equation: $s = 5.5w$, where s is how much you saved and w is the number of weeks.

Among the three of you, who was able to save the most amount of money each week, and who saved the least amount per week? Please explain how you got your answer.

Unit 3: Solving Equations and Systems of Equations

Activity 3.1

When bottom fishing, some fishermen use their arm length to estimate the lengths of their fishing lines. A normal bottom fishing set-up consists of taking your arm length and adding 6 inches to that. You then double this length to make 60 inches. Write out an algebraic equation to describe this, and use that equation to find how long your arm length is.

Activity 3.2

Among the fishing community, it is common to keep your favorite fishing spots a secret. This prevents overfishing at these spots. A couple of your friends have decided to shared their favorite spots in a bay by giving two equations each—their fishing spot is supposed to be at the intersection of the two equations. Find the intersections of the following equations (if you can!), and see if they were honest or lying in telling you their secret spot.

Friend	Equation 1	Equation 2	Solution	Lying or honest?
Lani	$y = 2x + 10$	$y = 10$		
Aloha	$y = 10x + 15$	$y = 10x - 10$		
Koa	$y = 5x - 2$	$y = \frac{15x - 6}{3}$		
Malu	$y = -6$	$x = -5$		

Below is the bay with a Cartesian plane. Check your answers by graphing the lines on the plane.

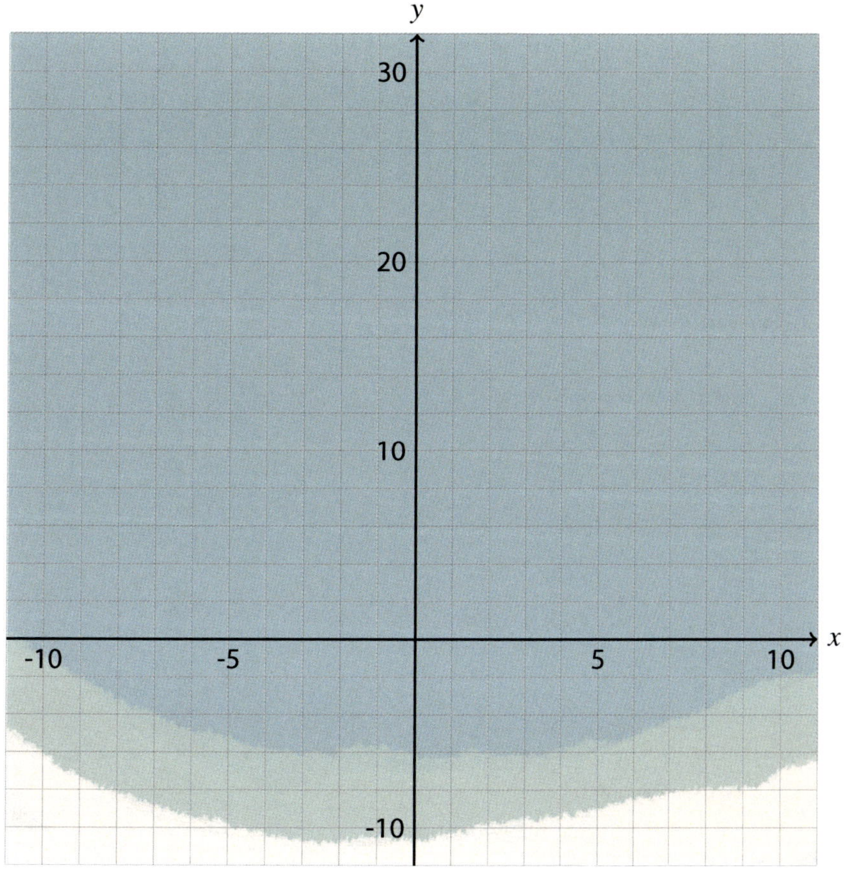

Activity 3.3

A modern GPS (Global Positioning System) can accurately pinpoint your location and track your progress throughout a day. Your GPS data shows that on your last two fishing trips, you traveled in straight lines and found a great fishing spot on both days at the point (-3, 2). To get to that point quickly, you decide to travel in a straight line from the shore (represented by the x-axis). Find *two* possible paths to get to your fishing spot (write the equations). Graph your paths and the point where you want to end up.

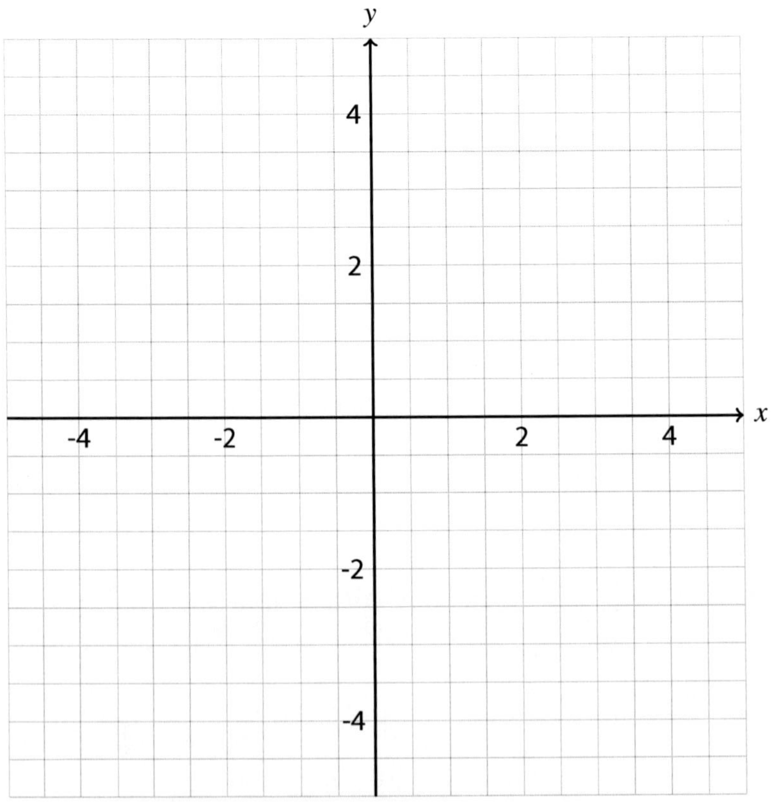

Did you just form a system of linear equations? How do you know?

Activity 3.4

Suppose that you want to buy fishing hooks online, and the shipping cost is always the same no matter how many hooks you buy at a time.

Last month, you bought 500 hooks for $13.00 (including shipping). This month, you bought 5000 hooks for $83.00 (including shipping).

What is the cost for each hook, not including the shipping charge?

What is the cost of the shipping?

What is the cost for n, number of hooks?

Activity 3.5

Your family is on a fishing trip, and you decide to catch something for dinner. Your favorite fish to eat is papio, which live in a channel that runs diagonally to the beach where you're standing. There's a particularly nice fishing spot in the channel, shown on the graph at point (4, -2). You want to cast your line from the beach straight to the fishing spot. You also want the slope of your fishing line's equation to be *greater than -1*. Graph two possible fishing lines.

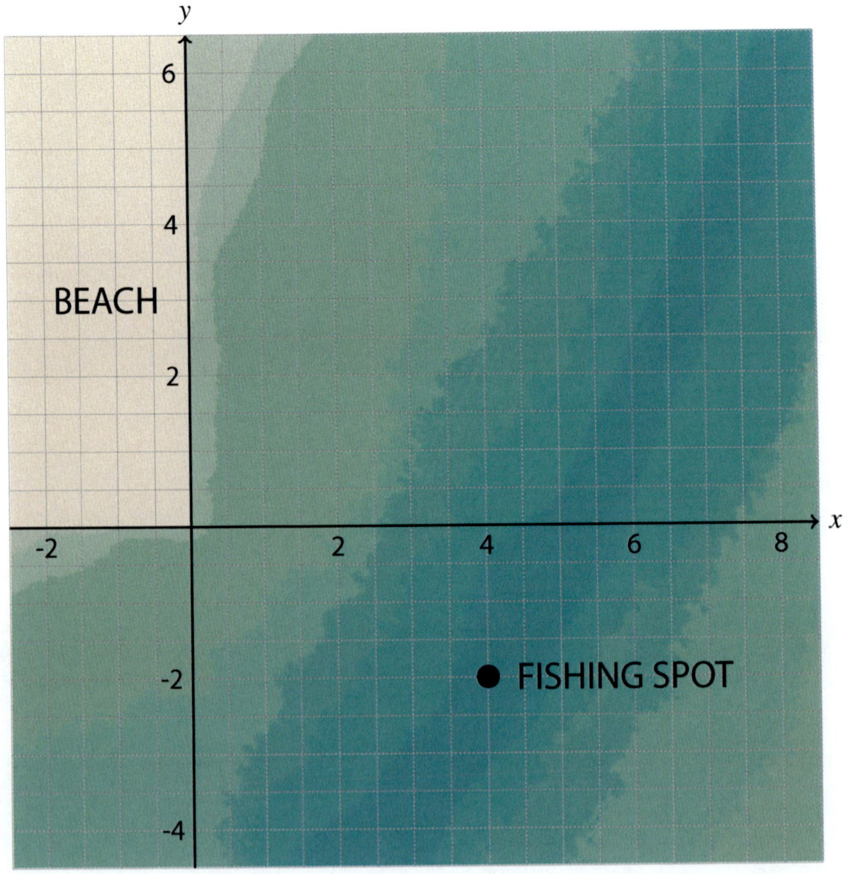

Unit 4: Transformational Geometry

Activity 4.1

Lay nets are a type of rectangular net with weights along one edge and floats on the opposite edge. When you lay it in the water, it stands up like a fence. You usually set up your lay net in a triangle like in the picture below.

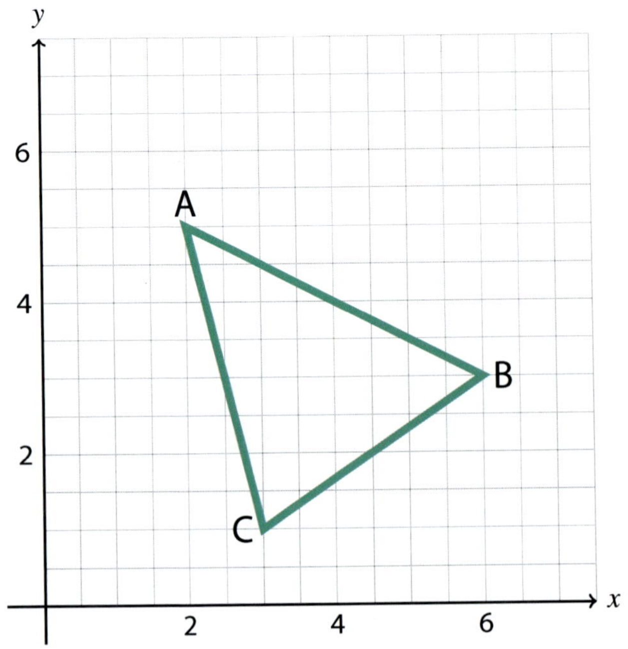

Your uncle gave you a larger lay net. Now that you have more netting material, you decide to take a look at your original setup and dilate it about the origin.

What could be a new possible location of point B in the bay? Please explain your reasoning. Is this the only possible position? Explain.

Activity 4.2

Lay nets are placed in triangles surrounding two different fishing holes in a bay.

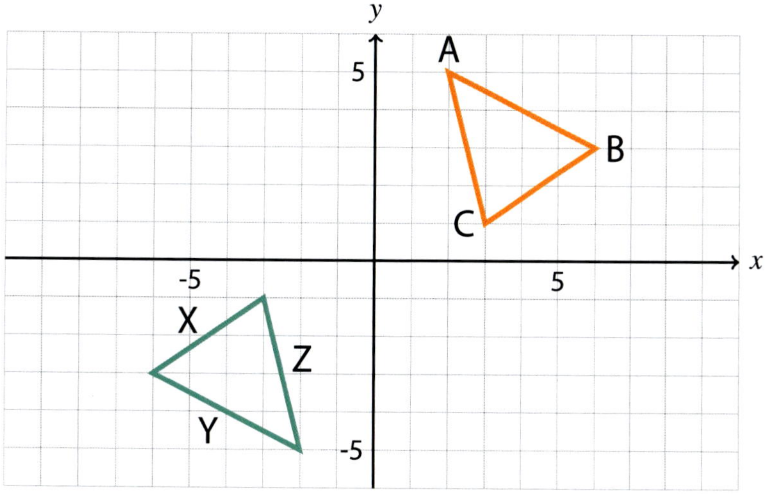

Check the appropriate boxes in the table to show which sides of the triangles have equal length.

	X	Y	Z
AB			
AC			

Activity 4.3

Rubbish piles floating in the ocean often attract small fish, which seek shelter under the rubbish. These small fish also attract larger predatory fish, which makes the rubbish piles good fishing locations. The shape on the graph below represents a rubbish pile. In a month, it will drift 4 miles south, and 5 miles west from its current location. Find the rubbish pile's new location and draw the rubbish pile on the graph.

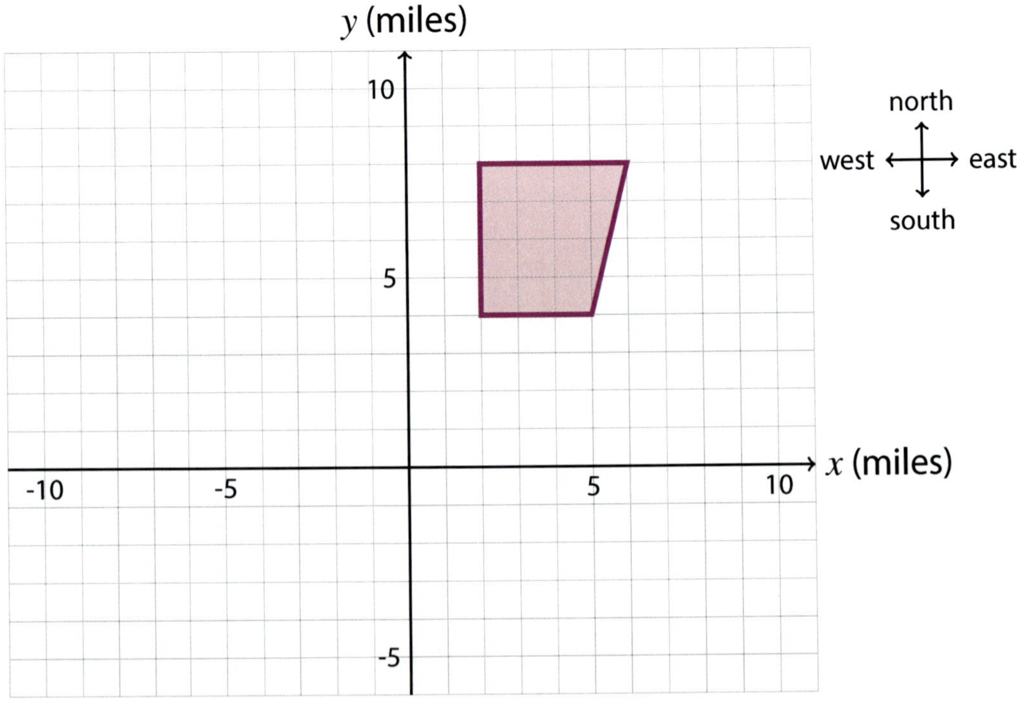

Activity 4.4

You noticed over several months that your fishing grounds have been changing due to the current and waves. In the first month, you placed your net as the blue triangle (ABC) below to surrounded a school of fish. In the second month, you had to move your net by a mathematical transformation. Finally, during the third month, you laid your net with the orange triangle (DEF) representing your net placement. Your net placements were similar triangles in all three months.

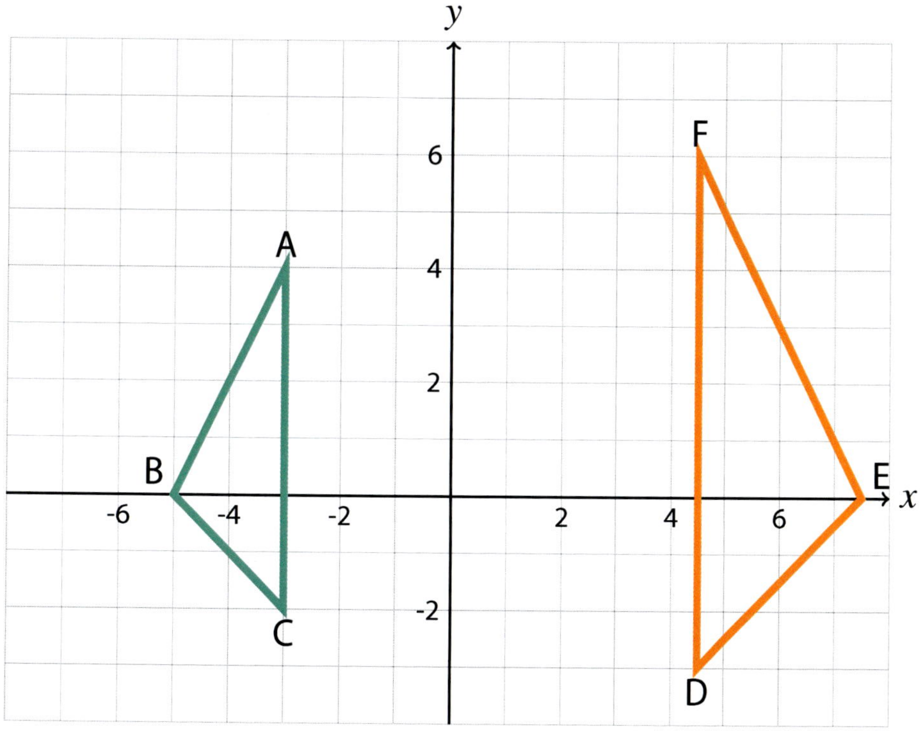

1. Create the second month's net placement by drawing a similar triangle to triangle ABC. Label this new triangle XYZ.

2. Describe how your second month's net layout could be obtained from triangle ABC with mathematical transformations.

3. Describe the mathematical transformation to obtain triangle XYZ from triangle FED.

4. Are the blue and orange triangles congruent? Explain your reasoning.

Activity 4.5

While designing your new canoe, you and your friend agree that the canoe should be symmetric when looking at it from the front.

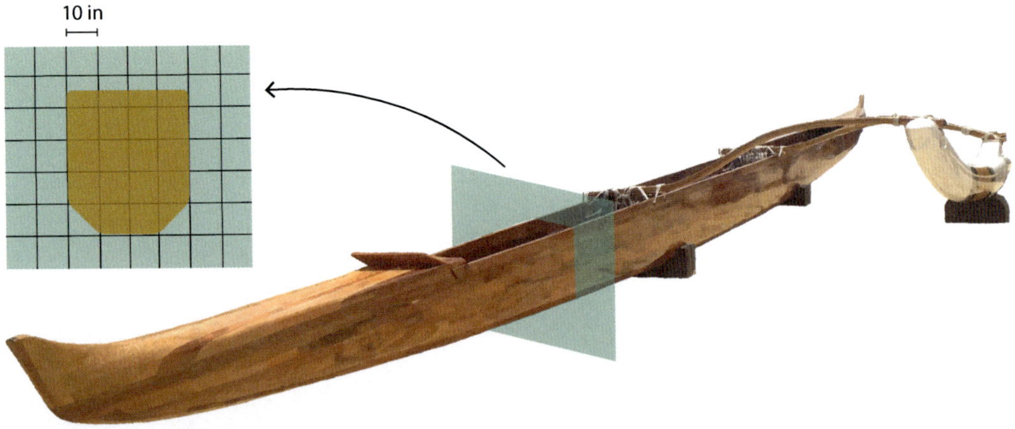

The two of you are sketching the design out on graph paper and discussing about the best way to make sure that it is symmetric. Then your friend says:

"If you reflect a polygon over the y-axis, then you just gotta keep the same y-coordinates, but take the opposite of the x-coordinates."

Is your friend's conjecture true?

If it is true, explain why the x and y coordinates become opposite/stay the same. You may also graph a few examples to make your explanation clearer. *If it is false*, provide an example.

Unit 5: Measurement Geometry

Activity 5.1

You want to lay a net in a straight line from where you are on the water's edge to a point that is 15 feet straight out and 8 feet to that position's right. How long a net do you need?

Activity 5.2

Ancient Hawaiians didn't use fishing reels—not even when catching *ulua* (a large predatory reef fish). Instead, Hawaiians used a traditional method was called *kau lā'au* "hang stick". Check out this YouTube video on Hawaiian ulua fishing: `http://bit.ly/2pTqTs2`.

Kau lā'au involves using a long stick with a rope at the end, which hangs a bait directly below it. Below is an image of a modern pole, represented by the line segment AB and its fishing line represented by BD. The pole used in the kau lā'au is represented by AC and it's line CD. Find the angles x, y, and z.

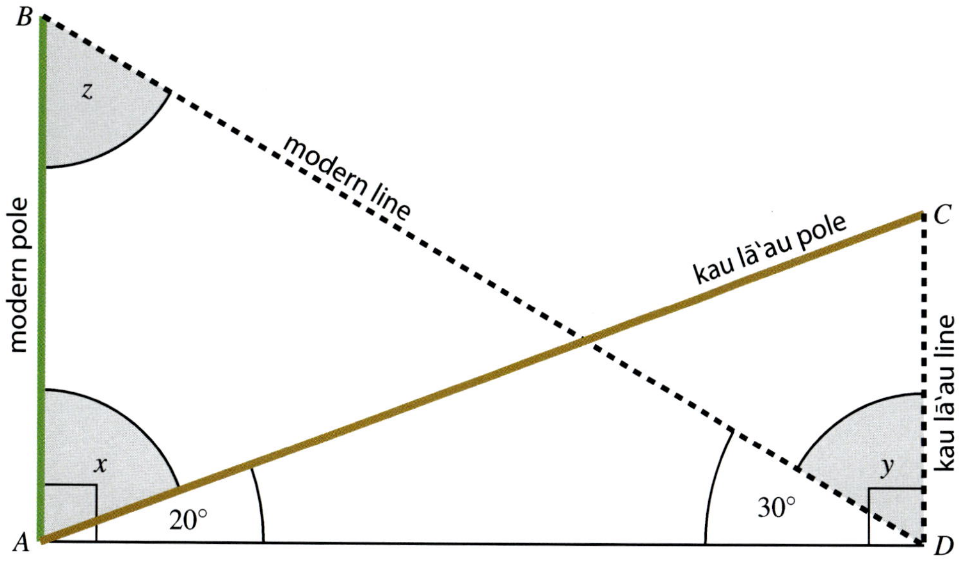

Activity 5.3

When constructing a speargun, it is easiest to start with a block of wood that is a rectangular prism, like a long shoe box. Every angle of a rectangular prism is perpendicular, and you can check your angles with a special tool called a *square*. If you do not have a square, then you can make one with Pythagorean triples. We just need to get 3 pieces of wood each cut to a length of a, b, and c, such that $a^2 + b^2 = c^2$.

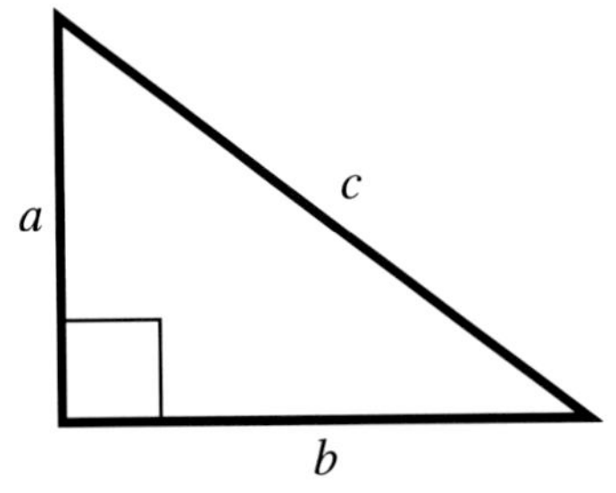

Find *two* sets of numbers (measurements) that our a, b, and c sticks might have to make our square tool.

Activity 5.4

1. Fishing gear takes up a lot of room and organization really helps with getting your fishing set up ready. One common item among all fishermen is a tackle box. It is a box used to organize fishing gear such as hooks, swivels, lead, etc. Say you wanted to make your own tackle box which has a square bottom, a height of 1 inch, and a volume of 64 cubic inches. What should the dimensions of your final tackle box be?

2. Say you wanted a tackle box for your lures, which are quite big, and you wanted your box to be a cube with a volume of 1,000 cubic inches. What should the dimensions of the lure tackle box be?

Activity 5.5

You are looking online to buy a new cooler for your fishing trip. The website you are looking at has the following description of their coolers:

"These new *cube* coolers are perfect for fishing trips!. Get your coolers with the following volumes: 1 ft^3, 64 in^3, 96 in^3, 125 in^3, 200 in^3, 256 in^3, 333 in^3, 361 in^3!"

You want a cooler that has *integer dimensions*. Sort the coolers out in the following table to help you decide which cooler to get.

A. Perfect square volumes (not perfect cubes)	B. Both perfect square and perfect cube volumes	C. Perfect cube volumes (not perfect squares)	D. Neither perfect cube nor perfect square volumes

Activity 5.6

It is illegal to catch papio that are too small. In order to figure out how long it takes papio to grow to a legal catching size, you decide to raise some baby papio. To raise baby papio, you need to make a *cylindrical* tank with a volume between 100 and 120 cubic feet. Find 3 different possible configurations that will give you your desired tank size.

	Tank #1	Tank #2	Tank #3
Radius (feet)			
Height (feet)			
Volume (feet^3)			

Activity 5.7

Native Hawaiians have always been excellent at building ponds for farming fish. In modern times, many societies raise fish in large cages or nets in ocean. When fish farming, it is important to consider the volumes of the cages. If we had a cone cage with a height of 27 feet and a base diameter of 32 feet but wanted a sphere cage with the *same volume*, then what should the radius of the sphere be?

Activity 5.8

Floaters are often used in fishing to keep baits and lures off of the ground. Some floaters float better than others and *how much something tends to float* is called *buoyancy*. The more volume an object has, the more buoyancy it also has. Suppose that you have two floaters: one is a sphere with radius of r, and the other is a right cylinder with a radius of r and a height of $\frac{4}{3}r$. Does the sphere have a greater volume, the cylinder have a greater volume, or do they both have the same volume?

Activity 5.9

While out on a fishing and camping trip with your friends, a storm began to approach. So you and your friends rush to make a frame for your shelter. Between you and your friends, you have sticks of lengths x, y, and z inches, such that $x < y < z$. You also notice that $x^2 + y^2 < z^2$.

One of your friends say that this is perfect and that you can create a right triangle to build part of your frame for your shelter. You, on the other hand, think that this is not correct and that you will have an obtuse triangle, which won't be good for your shelter since it won't be standing properly. Come up with a reasoning to justify who is right.

Unit 6: Statistics

Activity 6.1

The following table shows the amount and weights of catches of oama.

Number of oama caught	Weight of catch (oz)
10	24
15	30
12	23
18	42
22	47
17	34
19	39
25	58
14	26
11	18

Construct a scatter plot of the data in the table on the graph below:

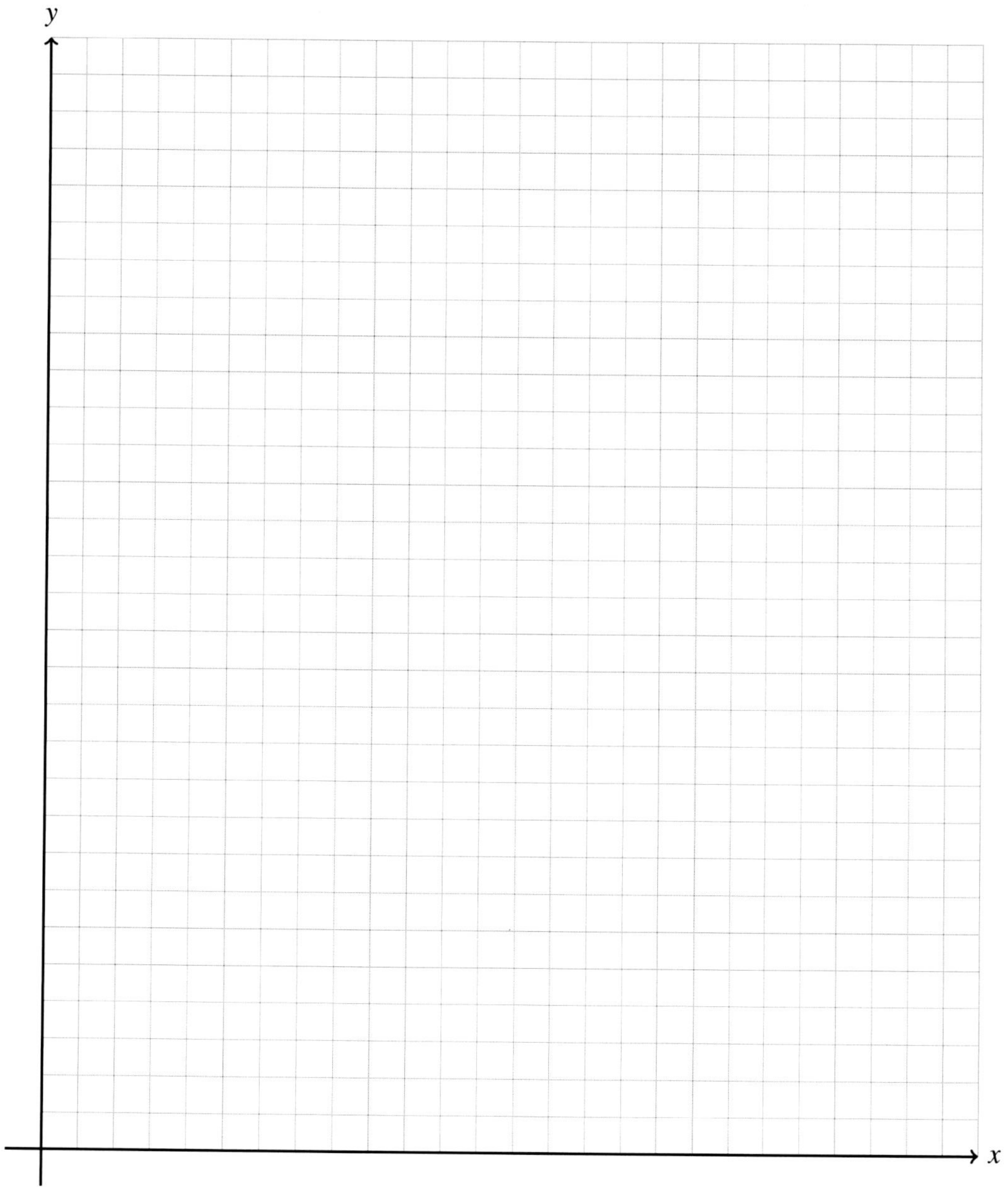

Draw a line of best fit on the graph for the data. Use this line to find how much more oama is needed to increase the weight of a catch by 40 more oz. Explain how you found your answer.

Activity 6.2

The price of fish changes a lot with "supply and demand." For example, in winter, ahi is normally not biting, so they're hard to catch and the *supply* is low. People want to eat more fish during the holiday season so the *demand* is high. So fresh ahi can get really expensive in the winter months. The "grade" or quality of the fish can also affect its price.

Let's assume that for each higher quality grade of fish, the price increased by 10%. Fill out the table below to create an example of how fish prices could depend on grade, with five star (★★★★★) fish being the best, and one star (★) being the worst. Start by choosing a cost for the one star (★) fish and work your way up.

Fish grade	Cost per pound of fish ($)
★	
★★	
★★★	
★★★★	
★★★★★	

After you have finished filling out the table, determine if prices are changing linearly with grade. Please explain why or why not.

Activity 6.3

Summer fishing trips are always a fun way to spend time with family and friends. As a reward for all your hard work during the school year, you can camp out, fish, and relax before heading back to school. Catching and eating fresh fish makes the experience all the more memorable. As much fun as fishing trips are, there can be a lot of planning that goes into it.

Let's plan a week-long fishing trip for your family of five (including you). You are asked to shop for delicious food and to pick a fun fishing location.

There's a store nearby that sells "meal packs." Three meal packs feed one person for one day.

Meal pack	Description	Price	Total cost for your family
I Stay Broke	Canned beans. Great for saving money!	$2.00	
Maka'āinana	Spam and rice. The classics!	$2.50	
Mouth Stay Broke	Kalua pig, salmon, and more. A local favorite!	$3.50	
Fat Like Wallet	Everything. Never go hungry or unsatisfied. Eat like an ali'i.	$5.00	

1. What are some interesting things that you see about the prices of the "meal packs"?

2. For each meal pack, find out the total cost(s) to feed your family for the week. Fill out the table above. Explain how you found your answer.

The following fishing spots are all around the island, and we have to consider the quality and cost of each location. Your family decides to take your family's truck to transport everything for the trip, and the truck gets *20 miles per gallon*. The gas price is *$3.25 per gallon*.

Here's some information about each fishing spot.

Fishing spot	Description	Distance away	Travel cost
1	Far and hardest to get to, but is full of fish and small wildlife. Secluded and very pretty.	76 miles	
2	Far, but has a good amount of fish. Clean and not too crowded.	43 miles	
3	Not too far. The fishing activity is okay. But a little crowded.	22 miles	
4	Closest spot, but doesn't have a lot of fish. Crowded, a little noisy, and a little dirty.	5 miles	

Keep in mind that the travel time will affect your fishing and camping time.

3. Calculate how much it will cost to travel to each of the different fishing locations. Fill out the previous table. Explain how you got your answer.

4. Which meal pack(s) and which fishing spot would you choose? Justify your answer and find the cost of your fishing trip.

Made in the USA
Monee, IL
02 November 2020

46558884R00033

Made in United States
North Haven, CT
22 April 2026

10363102R00031